The Arborists

Lonnie Ray Atkinson

The Arborists
Lonnie Ray Atkinson
(written in 2017)

Cover design by
Lonnie Ray Atkinson and Clayton Pollard
Cover Source Material: Ana Gram / Shutterstock.com

ISBN 979-8-9855830-4-5
UnsafeWords
unsafemedia.com

Contents

A matter of who you're trying to impress 5
Monsters and Saints 13
Dance with me 23
We have certain criteria 29
Having the same conversation 39
You better get used to waking up 47
We should have a kid 61
Belonging somewhere doesn't make you any less lost 71
Someone to feel connected to 81
It's not you 89
Come with me 95
Running from the sins of men 103
A morbid curiosity 117
I'm here because I killed God 131
Still capable of hope 139
A way back to that place 149

A matter of who you're trying to impress

Richard turned up the lantern and set it down between them, as Denton started to come to. When he could finally make out light from shadow, Denton saw Richard, a stranger, across from him, sitting on an old milking stool, a shotgun resting on his lap. He felt the rag in his mouth and the rope around him. He was fixed to the chair, his hands tied behind him. He could make out that he was in a barn. He could hear the rain and wind outside.

He sat still, and stared.

"Hello, Denton.

I know. Not exactly what you were expecting tonight.

Don't worry. I didn't come here to kill you. I'm just here to pass the time with you until this storm blows over."

Denton was careful of his movement as he tested the rope on his wrists. The stink on the rag became more pronounced. It made him conscious of his swallowing.

"I understand you're confused right now. And I imagine you're trying to figure out if you know me. But you don't. And truth be told, who I am doesn't even matter.

What matters is who you are. Or rather who you're not going to be after tonight.

Now, I'm gonna pull that rag out of your mouth, and we're gonna have a little talk. But before that happens, you need to understand. I don't have to hurt you. But if I decide to, nobody's gonna hear you hollering out here. So it's in your interest to stay calm and answer what I ask you. You ok with that?"

Denton nodded.

"All right then."

Richard walked over, stuck his finger between the rag and Denton's cheek, and wrested it from his mouth. Before he could get back over to the milking stool, Denton had already started.

"Boy, I don't know who you think you are. But if you don't let me go, you are *going* to die."

"Everybody dies, Denton."

"Yeah, but not everybody knows who's going to kill 'em.

I'm not sure what's going on here, but I believe you when you say you didn't come here to kill me. But if you ain't going to kill me, you best untie me. Because if I have to get loose myself, I am most definitely going to be the one to kill you."

"No, you're not. You're just talking big because you think that's what you're supposed to do in this situation. But you're wrong about that. In fact, Denton, you're wrong about a lot of things. I don't want to be here anymore than you do. But a consequence of me being here is that you get the chance to not be wrong, or at least not be wrong anymore. Now, I can put the gag back in your mouth and we can sit here in silence. But if we do that, I fear you're just going to keep finding yourself in the same wrong place. And that means I'm going to keep finding myself in this same wrong place."

"Who are you?"

"I'm just somebody trying to pass the time."

"You picked the wrong way to pass it."

"Maybe."

There was so much Richard wanted to say.

"But let me ask you. What good did you think you were going to do tonight?"

"I don't know what you're talking about."

"Yes, you do." The rain was getting harder outside. It forced Richard to raise his voice.

"Did you think you were going to make history tonight? Or did you think you were going to stop history from being made? Or did you just not think much about it?

Do you see yourself as an important man, Denton? Because you know there's a good chance tonight would have made you famous. Did you think about the newspapers tomorrow or next week, or the newspapers next year or ten years from now? Did you think about which picture they would use in the paper, what people would think when they saw your picture? Not tomorrow or next week, but next year or ten years from now. Did you think about what your picture might look like to people a hundred years from now? Did you think it through that far? Did you think you could see that far into the future to comprehend your actions here tonight?

Or did you just not think much about it?"

"Hmm."

"Something funny?"

"Yeah, all this here. All this here over a coon."

Richard steadied himself off the stool, made his way past the lantern, and drove the butt of the shotgun into Denton's stomach.

He let him gasp a while. Enough until he could be sure he was listening again.

"Now, that *man* you're referring to... is a man of nonviolence. I'm not.

So if you call that man one more name, I'm gonna pull off a couple rounds in you and then throw your body to the hogs."

Still trying to catch a solid breath, Denton looked at the ground as he spoke.

"Now who's trying to sound tough?

You think you really know something; sound like one of these college boys. But if you had any sense, I'd already be dead."

He looked up into Richard's face, trying his best to find his eyes.

"I know what you're asking me. You're asking me if I ever stopped to think I might be wrong. But, tell me something, college boy. You sure you're right?"

"I wouldn't be here otherwise."

"Me neither."

"But that's just it. One of us is wrong."

"Are you sure of that?"

Richard almost paused.
"We can't both be right."

"But we can both be wrong."

"What's that supposed to mean?"

For the first time, Denton felt he had the upper hand. "It means what I said.
We could both be wrong. What matters is if our gamble pays off."

"What gamble is that?"

"Well, the way I see it, you're gambling on who's really right and who's really wrong.
I'm gambling on who's going to be the one doing the judging."

"And you think that's morality?"

"It ain't about morals, boy. It's a matter of who you're trying to impress.

And I know who my people are. Do you know who your people are?"

Richard knew better than to answer.

"What I know is that I'm the one trying to prevent a lot of unnecessary suffering, and you're the one trying to allow it."

"Then I guess we'll see whose gamble pays off."

Richard hated how satisfied Denton appeared. He hated it, but he was careful not to show it.

"I guess so."

"You guess so. Hmm."

If not for the stink of the rag clothing his neck, it would have been amusing to him.

Still working on the rope around his left wrist, he decided to get him talking again.

"Why are you really here?"

Richard obliged, but not for the reason Denton assumed.

"I'm here because I need you. I need you to convince your pals that if you try planning something like this again, I'll know. And just like I did tonight, I'll show up again. And there ain't nothing you can do about that. Except next time, I'm going to kill you.

You and Jeffrey Dean and Bill and your deputy friend. And if I have to kill more than that, then that's what I'll do. I need you to understand that, Denton."

Denton replied honestly. "If it ain't us, it'll be somebody else. Sooner or later."

This time, Richard was the one almost amused. And sickened at the same time.

"Well, that is something we can agree on."

He raised his head up to listen.

"Storm's starting to let up, so I'm gonna go on ahead. The boy who slops the hogs can cut you loose in the morning."

It was too abrupt; Denton started to wrestle. Fighting at the ropes, he fell over in the dirt. He couldn't accept that he wouldn't get his chance.

Richard blew out the lantern, and threw the shotgun in the loft above them.

He stopped before walking out into the rain.

"End recording."

Monsters and Saints

They didn't always sit that way, chairs in a circle. But it did make the most sense. The walls were such an imperfect gray. If they weren't looking at each other, they'd be looking at the walls. No pictures or mirrors. No screens. No books in the corner. Just each other. Eight in all, including Richard, including Abbott.

They were there voluntarily. They had all they needed. Chairs and walls.

It was hour six. The focus had turned to Richard.

Clayton asked, sincerely, "I don't get you, Richard. If you're not going to kill them, why not just drug them and be done with it?"

Before Richard could answer, Ming Yan took the liberty. "Because he likes to get all philosophical with them."

Then Norman. "Don't you know? Richard's going to write a memoir."

"I'm not writing a memoir."

Clayton believed him. "Then what's with all these little talks of yours? What do you care what people like that think?"

"Because it's interesting. Don't you think it's interesting? Don't you think there should be some kind of record beyond just our logs and reports?"

"For what purpose?" Marwan was also sincere. "We're the only people that will ever get to see it. They're not going to get to see it, and it's more their history than ours."

"I don't know; I just think it's important. I guess I think there's something to be learned with all this, more than just hunting the technology."

Norman couldn't gauge whether it was time to be clever or to contribute something real. "It's like you're a historian for things that almost happened."

Not really correcting Norman, Rachel added, "Except it did happen before we got there."

"So he's an anti-historian." He had decided on clever.

"History my ass! You want to act like this is all some kind of movie or something." Bryant was always good at twisting it into a different conversation, or at least making the conversation about himself. "You think that when this is all over, you're going to be the star of the show just because you got all these recordings.

Except no one remembers who did what for the second-best film of the year. The only one they remember is best picture and best director. And those awards are going to me."

Before anyone could answer, Ming Yan took the liberty. "An award that no one's ever going to know about. You'd be lucky if they let you put it on your tombstone."

"I don't need it on my tombstone. It's enough that I know it, and that all of you know it."

Rachel could see, it was time to join in. "Yeah, maybe Bryant's right. Maybe we could have our own little awards show. It'll be two minutes long and we can have it at the beginning of Group, before Clayton starts in on how old he's feeling."

"You can joke all you want. But I'm dead serious. There's going to be a ceremony, and I am going to make an acceptance speech, even if it's just for you losers. I don't care if I have to haul a red carpet in here and make my own trophy; I'm basking in some glory."

Norman took his cue. He wished he was as clever as Bryant. "Why not? Maybe Richard can record it and it will go in the documentary they make about us after we're all dead."

"Or it can go in the pile with the rest of his recordings that no one is ever going to see." Ming Yan did not wish to be as clever as Bryant. "Seriously, there's nothing in the Agreement that says when or if they're going to reveal the Program."

Rachel had expected some interjection by this point. "What do you think, Abbott? You think they'll ever let the public know?"

"You all know what you agreed to. And you all know it doesn't matter which one of you finds the place. You were heroes the moment you signed up, heroes because you all agreed to be unsung. Sacrificing that fame for just knowing the good you're doing."

Having heard this over and over, Norman was a bit closer to no longer buying it. "That's assuming that any of us are ever going to get it right."

"Of course you're going to get it right. You're too good not to get it right."

"That or the law of averages."

The remark hung in the air, gathering affirmation. Uncomfortable with the course they were on, Rachel turned back to Richard.

"So did it work? Did you save him?"

"Yeah."

Bryant asked the obvious question. "How many times did it take?"

"Three."

"Three isn't bad. It took Norman seven." It had taken Rachel two.

"Yeah, but I should have listened to you."

"It's nothing to be ashamed of. We all saved him."

Ming Yan corrected, "All of us except Clayton."

"And there's a reason why I didn't save him. And it's the reason why you all found out that it didn't change anything."

"It changed a lot. Just not this." Marwan didn't like the implication.

"We're here to get a job done, not to be sentimental."

"And that means what?" Of course, Ming Yan knew. She just refused to be made ashamed.

"It means that you saved him not because you thought it would work, but because you wanted to see what he would have gone on to become or gone on to accomplish, even if it didn't change this."

Rachel rebutted, "That's not true."

"Then why did you? It's either that or you saved him because you just didn't want him to die like that. And don't mistake what I'm saying; I didn't want him to die like that either. But whether or not he dies like that or dies in his sleep on his hundredth birthday, this isn't about saving saints."

"Then, tell us, old man. What's it about?" Normally the contrarian of the Group, Bryant didn't care for the judgment either.

"It's about getting rid of monsters."

Rachel suspected Clayton was as sure as he let on. "Maybe. Or maybe what you're doing is no different than what you think we're doing. Who's to say you're not just curious? Maybe you're not sentimental. Maybe you're vengeful, towards the monsters. Except you can't see, that vengeance is rooted in sentimentality."

"I don't deny that I want to see what the world would look like without them. The difference between us is that, for me, it's not just a score being settled. I believe this is going to lead us there."

Marwan seconded Rachel's hunch. "And why are you so sure?"

"Because the monsters got us here. The less monsters you have, the less saints you need."

Bryant also didn't like how sure Clayton was, but not for the same reason as Rachel and Marwan. "You're just saying that because you think it's easier to spot the monsters."

Trying to dampen the implications, while still holding to his point, Clayton offered, "Maybe it is easier. But we're all supposed to be doing what we think is right. And I'm doing that."

"Except that doesn't matter either. Even if you are doing the right thing, it doesn't matter." Ming Yan was simply tired of the lecture. "It only matters if you know who the monsters are."

"And you think I don't?"

"I hope you do. But seeing that you haven't found the place either, assuming your strategy is the correct one, there's a possibility you don't."

"And what makes you think you know who the saints are?"

"What makes either of you think you know?" Bryant saw an opening to regain supreme contrarian status. "Who's who depends on who wins. It depends on who gets to write the story. We should all know that by now."

"Maybe so. But then the question becomes: Which is it more dangerous to be wrong about?

If I'm wrong about the monster, I still cause less harm than if you're wrong about the saint."

Bored with all the philosophy, Norman argued disingenuously, "What does it matter? Until we find the place, being wrong doesn't mean anything. The worst thing that happens is that we have to come into work the next day."

"But it does matter if they're wrong. We have to figure these things out or we're going to be coming in here forever." Norman could do what he wanted, but Marwan wasn't going to let everyone else give up. Not this time, not this easy. "My question is: What if you're both wrong? What if neither exist? Or what if they only exist in the places we found them? There could be another place where they switch roles altogether."

"The monsters become the saints?" Bryant kind of liked the idea.

"And vice versa," added Rachel, though she kind of didn't like that one.

"That's right. What if we're all wrong about the way we're looking at it? What if it's not about who's who or that one moment in time? What if it's about someone we'll never even know existed or something so small we can't even know it happened?"

"Or what if it's about millions of those things?" Ming Yan heaped on.

"That's what I'm saying. Our job is trying to find things to focus on.

But what if that's the problem? What if that's the reason we haven't been successful? What if it's too big to focus in on? What if it's too many things? What if we're being asked to guess some random number, and the only number we're willing to wager is our favorite?"

"Or what if we don't even know what all the numbers are?" More heaping on.

"Exactly."

Bryant began to laugh. "But it can't be that! That would make our jobs even harder."

The remark hung in the air, gathering affirmation.

Dance with me

He'd been standing there for over a minute, since the scan downstairs had signaled arrival. He looked at the door, then the corners framing the door. Then the ceiling connecting the corners, then the floor.

But he wasn't really looking. He had just recently gotten to the point where he didn't pay attention to the architecture or the scheme of colors. Muted in the kitchen and the bathroom, yet pardoned in all other rooms by a far more attractive gray.

It was no longer fancy. Now it was just a furnished apartment. Clean and secure.

The door unlocked. Annasophia walked through it.

"I'm sorry I'm late. Are you ready to go?"

"I am ready to go. Buuuut firrrst, you have to…"

Richard turned down the lights and pressed play.

"Dance with me."

"What!?!" Annasophia whispered her surprise.

He wrapped his arm underneath her elbow, gently. With just enough rhythm.

She complied, squinting sweetly, wondering what he was up to.

Then it hit her.

"Heyyy. How did you know I liked this song?"

"We heard it back when we were dating, when we were on our way to have dinner at your sister's. You stopped and started crying."

"I did?"

"When we saw that little old man playing piano on the street. This was the song."

Annasophia brought her head slightly back. "You're right. We were walking it and took that shortcut because we were already late, but I still wanted to stop."

"You stood still and closed your eyes until the end of the song, and then gave the old man a hug."

"I'd never heard it played on a piano like that.

You've got a good memory. I forgot all about that."

"You had to wipe your eyes and collect yourself before you could say anything. I figured it must have meant something to you."

Just hearing him describe it. "It's the only song I ever saw my grandparents dance to."

"Yeah?"

"It was at my cousin's wedding. I think I was like nine. My grandmother could barely walk by then. Poor thing was all hunched over. But when that song came on, she looked up at my grandfather and he walked over and took her hand. And she stood straighter than I had ever seen her. And they moved out onto the dance floor and swayed together like they were teenagers at the prom.

I can't even remember seeing her walk by herself after that. Watching her become that person again showed me how magical even the first few notes of a song could be."

Richard pulled back so she could see his eyes. "Watching you listen to that song is one of the moments I realized that I was going to fall in love with you."

He then pulled her back in. They knew each other enough to leave the rest to dancing.

When the song ended, Annasophia kissed Richard. The moment deserved it, as did he. Plus, she needed to break the spell if she was going to ask.

"So what's up?"

"What do you mean 'what's up'?"

"Tonight, an old song. Last night, my favorite meal.

The night before that, you suggest we go visit my mom. Are you getting ready to tell me something terrible? Have you had another wife this whole time?"

"Two wives actually. You're the middle one."

"I knew it. But I told myself the other wedding albums in the closet were nothing."

"You really should have paid more attention to the signs."

"No but, really, what's going on with you?"

"Can't I just do nice things for the person I love?"

"Oh, you can; I'm just not used to this much sustained romance."

"I don't know. I guess with everything going on lately, it's got me to thinking about all the times I've missed telling you what I should have told you or doing something I should have done."

"It's not like I'm always on top of my game either."

"You give me every reason to do everything I'm doing. Being able to come home to you…

I just don't want to miss any more moments with you."

He could barely finish his sentence. It surprised her, the sudden emotion.

"It's ok."

"I love you."

"I love you too."

Richard pulled her in, again, and led her to swaying. Dancing to a shared feeling.

They danced a little longer, and then Richard conceded.

"Speaking of moments, we should probably get going."

We have certain criteria

It was hard to get his mind off where he was. Everything about the office, even the gray on the walls, spoke to it. The whole building, all the buildings. Clean and secure.

It was easy to keep humble.

Still, Richard could sense traction, even if he was afraid to believe it.

There was no way Galya was a pushover. That was obvious the moment they sat down. Maybe she was testing him in a way he couldn't recognize. She was, after all, the smartest person in the whole building. Maybe all the buildings.

Why wasn't he more intimidated? Why hadn't he fallen apart?

Could he really be doing this well?

She continued. "We call them coordinates."

"Like a GPS?" Oh no. Worry set in immediately. Was that too simple?

"Kind of like a GPS, but with additional, for lack of a better word, dimensions factored in. Think of reality as a tree, but a tree without any branches. Your job would be to go back to a certain point on that tree and create a branch.

And once that branch exists, once that reality exists, its coordinates exist.

We can then send you to any point of that reality's chronology up to the present time."

"Why only the present; why not the future?"

"The coordinates don't exist yet, at least not in the way we have the capacity to capture them.

We can go sideways and backwards. We just can't go forward."

"So what happens once I create a branch?" Was it too presumptive to use "I"? She'd been using "you" and "your." It must be ok.

"We send you to our present moment in that reality, and you see how your branch turned out."

"You mean if it turned out better."

"Correct. We have certain criteria we're looking for."

"That's the technology part."

"Correct. We know that both the momentum and direction of technological development depends on various key factors. Your job would be to create a branch of reality with circumstances conducive to developing the kind of technology that we don't have."

"You mean the kind of technology that we don't have yet."

"Correct. Given enough time, certain discoveries are inevitable."

"But we don't know if we have that much time."

"Correct."

Richard leaned back in his chair.

"Wow."

It was a lot to be correct about.

For the first time in the interview, Galya smiled.

It was sympathetic. Like she knew how much he needed it.

"We understand that this is a lot to take in."

"You mean being interviewed for one of the government's geo-engineering programs, even though I have zero science credentials? And then being informed we have a technology no one else knows exists?

Yeah, that's definitely a lot to take in."

"Any time you need, just let me know if there's something you'd like me to go back over."

"No, I get it. The way you've explained it, it makes perfect sense. Why work on coming up with long shot technologies that are not certain to work - and could have even more disastrous unforeseen side effects, when you could just copy what already worked for another branch of reality?"

"Correct. The problem is that there are multiple other divisions within our geo-engineering unit working on those types of long shots. It's my job to find the right people who can find the right branch before we have to employ any more of those options."

"So why me?" It was a risky question, but the time was right. "Like I said, I don't have any science background, beyond just what was required in school."

"We're not looking for a scientist. We're also not looking for some kind of time traveling action hero. We're looking for someone with the kind of analytical skills and vision to see the missed potential in each world."

"What makes you think I'm that person?"

"You wouldn't have made it this far if we weren't sure."

"But what if you're wrong, about me?" Richard was starting to feel cocky.

"We don't believe we're wrong. We've considered people with all kinds of backgrounds, and you fit a profile that we believe would be by far the most effective. You scored higher on parts of the test than anyone we've brought in."

Richard smiled, the one he normally reserved for Annasophia. "Only *parts* of the test?"

"There were only certain parts that mattered." Galya did not reciprocate. "Just the fact that you're asking if you're right for this job shows me how you think. It shows that you're more worried about the success of a program you just learned about than you are getting a job that could vastly improve the quality of your life."

"That's because I understand the project's importance."

To this, Galya reacted with a laugh. Like a treat to a dog, it was small. It wasn't even audible. Just breath, and eyes closing longer than a blink.

"What?"

"Either you're saying all the right things because our testing procedure is proficient, or you're playing me."

"To be honest with you, I'm asking all this because it sounds intimidating. What happens if the branch of reality you create doesn't meet those criteria? Then what?"

"Then you have some reading to do. We're looking for someone who can do a historical survey of that reality, of any given reality, and to then locate a specific moment in that chronology. Something you believe would make the most beneficial impact, if changed."

"You mean make a new branch."

"Correct."

"And if that new branch doesn't meet the criteria?"

"Then you find another branch to make."

"And another and another until I find the right one."

"Correct."

"What if I can't find a viable branch off of the original branch?"

"If you feel you've come to a dead end, you start with a new branch."

Richard rubbed his toes together inside his shoes to distract him from the racing of his heart, to keep from acting as surprised as he should be.

"You mentioned that I would be brought onto a team of people all doing the same thing. Would we be sharing branches?"

"No, you'll be working alone for the most part. The only thing you will share with your team members, aside from comradery, will be information.

Once you create a branch, it's yours. You can consider it your exclusive territory - to make of it what you will."

"You make it sound so grand." Richard said this, then immediately worried he had overplayed his cool.

"That's because it is.

You'll find that not only is this project vital to our survival, it's intoxicating. Think about what we're offering you. As you said, you have no scientific training, yet you could provide us a proven technology that will save our people. You can give us that. You can save billions of lives.

Make no mistake; this will be work. As the early team members will share with you, this will be the most taxing thing you've ever done. But it will also be the most enchanting.

Each of you are explorers of a new realm. You're to discover the possibilities of humanity. You're to create those possibilities."

"And yet no one will ever know."

"That's right. That's part of the Agreement.

We have an ethical obligation to limit who is aware of this technology. There are far too many wrong hands it could fall into.

Your reward will not be who knows about this. Your reward is being a part of it. By agreeing to be part of this team, you can know that every team member's accomplishments are yours as well. By simply participating, you will always own a part of what comes out of this program."

It was the first time in the interview Richard didn't know what to say.

Galya stayed quiet, and waited.

Having the same conversation

The walls were thick enough for them to yell. And they had decorated enough to prevent an echo.

But the gray. So attractive it was annoying. Clean and secure. As if any of that was true.

Annasophia was standing, as Richard sat there.

"Listen to me. Richard, listen to me. How long do you think I'm going to keep doing this?"

"Doing what?"

"This. This coming home to a stranger."

"What are you talking about?"

"You *know* what I'm talking about. You come in, take a shower, play immersion games for a couple hours, and then you go to bed. And that's it. You hardly eat anything. You're completely distant. You won't talk to me unless I force it out of you."

Richard hated that she had made it into a list.

"I told you I'm just going through some stuff at work."

"Then talk to me."

"You know I can't tell you anything."

"You can't or you won't?"

"I can't. And even if I did, you wouldn't understand."

"Then tell me." She was pleading. "You don't have to tell me specifics. But at least talk to me."

"There's nothing to talk about. I'm having a hard time at work right now. It won't last forever. I'm going to get through it, and then everything will be ok again." He could tell how hollow it sounded. And he knew Annasophia wasn't stupid.

"Will it?

I've been trying to have this same conversation with you in different ways for the last two months. But now I'm not sure if I even know why I'm trying.

It's like I don't know who you are anymore. Or it's like I do know who you are, and I don't like who you are."

"Are you kidding me?" Richard took the cue. "You don't like who I am? You don't like who I am?

Let me tell you something; you have no idea what I'm going through. You have no idea what I have to do so you can have this nice little life of yours in the Zones."

"Don't you do that. Don't you try and put this on me.

You think I care about any of that? You think I care about living in the Zones? I'll pack up my things and we can move back to that dump we had in Section 14. I'll do it tonight. If it means you acting like a normal human being, I'll go live in the Tunnels."

"A normal human being?"

"Yes, a normal human being.

Baby, I know you can't see it, but there is something wrong going on with you, and I'm just..."

"Quit telling me there's something wrong with me. Quit telling me that. Yes, I know I've changed since we got here. But you don't think you've changed too?"

"No, I haven't. Not in the last two months."

"Well, I'm sorry then. I'm sorry."

"Are you sorry, or are you just saying whatever you think I want to hear so I'll leave you alone and you can go back to distracting yourself?"

"What do you want from me? What do you want? I don't know what you want me to be. I'm going in - and I'm doing so much - and I'm doing the best I can. And all I hear from you is that I'm not good enough."

"It's not that you're not good enough. It's that you're putting everything in at work. But when it comes to us, you're not trying.

Baby, whatever is going on with you I don't think it's your fault. And I'm willing to work with you. I'm willing to go to counseling with you. I'm willing to hold you when you wake up at night freaking out. I'm willing to deal with a lot. But you have to try."

"I am trying."

"You're not trying. You're learning how to live with me and without me at the same time."

Her words reminded him of why he was so drawn to her in those early days. She was someone too smart to be dismissed or fall for lazy excuses.

"I am trying. I just don't know what you want."

"I want you to come back to me."

"I haven't gone anywhere!"

This is what Annasophia had feared. She loved him too much to let him get away with trying to rely on her love.

"So this is all I'm going to get?

You're just going to act like nothing's really wrong, and that it will all be ok in a little bit?"

"You're making too big a deal out of this."

"I thought you'd say that."

He hoped she was just walking away. He was even ok with her slamming things in the other room.

When she came back in, he looked up, and regretted all of it.

The bag might as well have been a gun.

"What are you doing?"

"I'm going to stay with my sister."

"What are you talking about?" Richard got up and ran over toward the door, between her and the door.

Unimpressed with his desperation, Annasophia stopped and stared at Richard. "I told you I can't keep doing this."

"Doing what? We're having a conversation."

"Yes, we're having the same conversation. And I can't keep having the same conversation."

"Your sister lives in Section 03."

"Yeah, and so did I when we met."

"This is ridiculous. Put your bag down and let's talk about this."

"We just talked about it, and you told me everything I needed to know."

"So what, are you leaving me?"

"Of course I'm not leaving you. But I can't make you do something you don't want to do. And I don't know how to get you to deal with any of this. So I'm going to give you the time you need to figure it out."

She then looked at him, allowing him to get out of the way.

"I can't believe you're doing this."

"I can't believe you didn't see it coming."

It hurt her that she was right. But not as much as it hurt him. And not for the same reason.

Richard took a few seconds, then moved out of the way.

You better get used to waking up

From every other room he had been in inside the Zones, it surprised him that this would be the least impressive. It looked like a room rented out for AA meetings. Something you couldn't make *your own.* Not like you would want to anyway. With that gray.

Just walls and chairs. The minimum needed for refuge.

Abbott politely cleared his throat as he closed the door behind them.

"Everyone, this is Richard. Richard, this is the Group."

Norman was excited it was another guy. He wondered if he would be as cool as the other new guy. "Is this our final addition?"

"From what I understand."

"Well, welcome aboard."

"Yes, welcome." Marwan's greeting was warm, though he couldn't betray his slight sense of worry.

Abbott moved on, sticking to business, as if he had somewhere to be. Even though he didn't. "Each member of the Group has the same job as you. Members use this time to give each other tips, as well as to create community."

"What he means is therapy."

"Decompression is the term I prefer."

"It's therapy," Rachel whispered.

Acting unbothered, Abbott picked back up. "These are the only people, aside from those in East Wing, who know anything about the Program. This is where you will be able to share and learn and make sense of everything."

Richard looked around at how little there was in the room, then at the others.

"I facilitate the sessions."

"He referees the sessions." Norman liked that he was the first person to give Abbott a jab in front of the new guy.

"What they mean is to expect a lot of trash talk."

"I didn't know this was a competition."

"What's the fun in just saving the world?" Like usual, Ming Yan spoke from a place of irony.

"Remember, everyone, we are a team.
That's why Group is important."

"Did you tell him what we call ourselves?" Norman loved this part.

"It's your name. You tell him."

"We originally thought about calling ourselves 'the Gods,' but the Program wouldn't allow it."

"So we unofficially call ourselves 'the Gods'." Clayton figured stealing Norman's thunder would be a suitable first impression.

"But officially…"

Norman jumped back in, "Officially, we settled on 'the arborists'."

"The arborists."

"Yeah, you know, tending to the branches."

"Ahhh." Richard nodded and smiled.

Rachel reciprocated. "In case you haven't noticed, we weren't chosen for our wit."

Abbott seconded, "Speaking of wit, expect a lot of *Quantum Leap* jokes, a lot of *Back to the Future* jokes, a lot of *Terminator* jokes."

"Except we're like bizarro terminators." Norman had come up with that one on his own.

Rachel elaborated, "Basically half our time is spent on arguing how we compare with old science-fiction pop culture references."

"Group lasts up to 7 hours, or as long as you want to stay."

"Seven hours seems like a long time."

"After a week and a half out there, these 7 hours are going to fly by." Marwan spoke softly. He remembered how he had felt in the beginning.

Clayton added, not caring about how anyone felt, "We have to fill time before we go home and fool our partners into thinking we…"

"That's enough," Abbott calmly interjected.

Abbott sat down in one of the two remaining empty chairs and motioned for Richard to take the other.

"Just so you know, you're not limited to the room for 7 hours. There are other facilities you have access to during this time."

"We just end up spending most of our time in here." Rachel knew having it come from one of the Group would be more reassuring than coming from Abbott.

"So what do you do for 7 hours?"

"We compare notes, give each other advice." It wasn't a lie. Marwan just preferred to focus on the more productive aspects of their time together.

Again, Clayton didn't care. "We fight."

"Yeah, a lot of fighting." Although Bryant had only just arrived himself, he was learning quickly.

"Some fighting. But mostly it's just a lot of storytelling." Again, not a lie. "What's the strangest thing you saw on your branch, getting out of difficult situat…"

"Sexual encounters." Norman waited gleefully for the eye rolls.

Rachel was first. "Here we go."

"Yes, please, let's save that for another time. I'd like to keep the focus on our new Group member, if that's ok with everyone."

Norman raised his palms. No need to rile up Abbott. Besides, he had already gotten what he wanted.

"Because the project is the first of its kind, there are no studies or textbooks to help us along. Most of the research that will come out of this project is based on the interactions you will have here in Group."

Rachel translated. "What that means for us is that because the Program doesn't allow you to talk about this with anyone else, this is who you can talk about it with."

Anticipating some snarky comment, Marwan quickly added, "Despite the trash talk and the fighting, which you will definitely play part in, this is a family. We understand each other on a different level."

"Enough with all the sappy stuff, just introduce everyone." Clayton was hard to read. Richard couldn't tell whether his gruff demeanor was real or for effect.

"Ok. I was getting to that.

As I said before, everyone. This is Richard.

Richard, this is Marwan and Ming Yan. They are our most senior Group members."

Ming Yan smiled at Marwan. "Veterans of the game."

"I thought this is a new project."

"Technically, the Program is only a few weeks old, but you could be out on a visit for as long as a month or more." It was the second time Marwan had mentioned this.

"That's right. These two have logged a total of fifteen and twenty months respectively."

"He means we've aged fifteen and twenty months within a little over seven weeks." For this, Ming Yan wasn't smiling. "Respectively."

"Thank you for the clarification, Ming Yan." Abbott proceeded, "Of course, Bryant you met at briefing and orientation. Bryant came on a few days ago."

Richard and Bryant both nodded, acknowledging their familiarity.

"Norman and Clayton came on two weeks ago.

And Rachel came on a couple weeks before them."

"I thought you brought on two at a time. Why is there only seven?"

Ming Yan had been waiting. "Because Galya's been in this world long enough to forget what it's like in that world."

Clayton affirmed, "The Program is smart enough to create a time machine but not smart enough to factor in history."

"Does that mean you lost one?"

"We lost three." Marwan felt like either he or Ming Yan should be the one to say it.

"Three?"

"Yeah, three. Devin Morris, Luz Manriquez, and Juanita Halliday." Ming Yan said their names just as much for Abbott to hear.

"That's why you made the cut. That's the--"

"What she means is that we've had to adjust our candidate criteria based on the results we were receiving."

"Depending upon how far you go back for a branch, it's far more difficult to move freely or unmolested without a certain amount of privilege." Marwan was clearly the more diplomatic of the two.

Ming Yan said the rest. "So they thought it would be more effective to discriminate."

"We thought it would be safer to…"

"Excuse me, safer to discriminate."

"We lost three people." Abbott conceded, matter-of-factly.

"No, *we* lost three people. You didn't lose anything.

And, besides, they knew what they were getting into."

"Ok. We've addressed this before, and you know I have no control over these decisions. But I will convey your views.

Right now, I think it's best we focus on welcoming our new--"

"And I think it's best that we let him know the good, the bad, and the ugly."

"Then maybe we could focus a little more on the good."

"Yeah, everybody. Let's focus on the good." It was hard to tell whether Clayton was being sarcastic.

This was Norman's specialty. "The good is when the bad doesn't happen anyway."

"I don't get it."

"You'll see pretty quick that almost nothing works out the way you think it's going to. So a lot of the time, the good is when you get lucky and things change in a different way than you expected, but it's still better than before."

"And that happens a lot," concurred Rachel.

"And that's good?"

"Yeah, it's good." Norman concluded, "But it also reminds you how little you know."

Richard took a moment to contemplate.
"So what's the bad?"

This one was Ming Yan's. "The bad is when you do something you're sure is going to work, but instead it makes it worse."

"Way worse." Another thing Bryant had come to know within his short amount of time.

"Does that happen a lot?"

"None of us are working on our original branches, if that tells you anything." Clayton took no pleasure in these words.

For the same reason, it shocked Richard. It shouldn't have, but it did.

He had just realized that this was the real orientation.

"What's the ugly?"

Now it was Marwan's turn. "The ugly is getting your heart broken when you've done something so big and then find out how little has changed."

Unable to relate just yet, but keenly reading the room, Richard feigned a few seconds of absorption.

He decided to ask something lighter.

"What advice do you have for me?"

This time, Rachel was up. It was the same thing she had told Bryant, and it seemed to be well approved of by everyone else. "You better get used to waking up."

"I don't know what that means."

"What we do is like stepping into the craziest, most lucid dream you've ever had. And it's one of those dreams where you can live for what seems like weeks, but then you wake up and realize you've only been asleep for a few minutes."

"I love those kinds of dreams."

"Yeah, but what do you normally do when you wake up from one?"

"I try and go back to sleep so I can get back to the dream."

"And how often are you able to do that?"

"Not often. And when I can, it's not the same."

"But you keep trying, don't you? You keep trying until the door has closed. And you're left awake, resenting your inability. Wanting that world more than this one."

We should have a kid

Even on the 18th floor, you could hear everything, outside. The noise was unpredictable. Sometimes frightening. It made the rounds of the trains the most calming minutes of the night.

Annasophia lay on her back; only one of them was afforded the luxury. They called it their Bob Marley bed; it seemed romantic. The truth was no one had anything more than a single in the Stretch. The rooms were just too small.

They had kept the furniture left by the previous tenants, just like they kept the eccentric scheme of colors left, different in each room. The colors were from someone else, someone they would never know. Colors so eccentric that, when they moved in, they made up a story about a little old couple who were artists but would only show their work to each other, never anyone on the outside. They would use the walls as a mixing palette, and when one of them would stumble upon a color so perfect, a color they knew no one would ever use for a wall in any civilized home, they painted the entire room that color. When the wife got sick, at 92, the husband, 90, collected all their paintings and rolled the canvases together into bundles, with only the backs showing. Little by little he brought the paintings downstairs into the courtyard and gave them to the different groups warming their hands around the barrels. He told them that he hoped there would be enough to keep them warm through the night. Then he went back upstairs, climbed in next to his wife in their single bed, and they kissed. Then went to sleep. Then never woke up. The only legacy of their art, walls left colors no *civilized* home would ever bear.

Annasophia looked at the chips in the paint, the walls scarred by who knows how many couples since then, families since then. She had no idea how much she would miss those walls, those colors, within a mere two years.

She had been lying there long enough that she wasn't really looking at anything.

"Are you asleep yet?"

Richard consciously included a bit of rasp in his voice. "If I don't answer, will you wait until the morning to bug me?"

"We should have a kid."

"Ok, that came out of nowhere." The rasp was gone.

"You didn't think it was ever going to get brought up?"

"It's not that I didn't think it would ever get brought up." Still on his side, Richard turned his head, as if talking to the ceiling. "I just wasn't prepared to talk about it tonight."

"Well, when were you going to be prepared to talk about it?"

"I don't know."

"How do you not know?" Annasophia sat up.

"I don't know. I don't know if I feel prepared to have the conversation because I don't know if I even feel prepared to be a parent."

"Right now, or at all?"

"I don't know. Whenever I have thought about it, the pros and cons seem out of whack. It's like the same positives are there that have always been there, but on the other side there's more negatives every day.

And anyway, we don't have the credits to do it right now even if we wanted to."

"People have kids on way less than what we have."

"Yeah, they have them. But what kind of lives are they going to have? We're lucky to be living in the Stretch. If we have a kid and the opportunity queue slows down, who knows where we'd end up?"

"You just don't want to have kids."

"I didn't say that." Richard finally sat up. "But if we are talking about it, then I have to be honest. Sometimes I think to myself that I don't think this is for children."

"Come on, this isn't that bad."

"I'm not talking about the Stretch. I'm talking about this life. Would you want to be a kid right now?"

"Children are as happy as they've ever been. It's relative to what they know."

"Yeah, but children grow up."

"Meaning?"

"Meaning we've grown up used to getting randomly displaced. In thirty years, everybody outside the Perimeter is going to be virtually nomadic.

And if they have a family, their kids won't even understand the concept of having a home."

"Yeah, but we don't know that." Annasophia said this, but didn't really mean it. "They could still find a fix."

"They. Who is they? Do we even know anymore?

Yeah, they could still fix this. Or they could keep failing, and, in a hundred years, two-thirds of the world ends up scavenging. They didn't even see what was coming twenty years ago. You think they know what kind of storms are coming in the next twenty?

They're not prepared for any of this. Or they're not preparing us. If it's not happening outside the Zones or the Perimeter, it might as well not be happening."

"So if it's *that* bad, then maybe we should go jump off a bridge."

"I'm not saying that."

"Then what are you saying?"

"I'm saying I'm scared.

You know how my parents died. And that was then. I can't stand the idea of our kid growing up without us."

"Just because it happened to them doesn't mean it's going to happen to us."

"Yeah, but just the chance of that happening is too hard for me to think about. And there's a lot more chances now."

"Yeah, but that's life in general. Anything could happen."

"I know anything could happen, but I feel like having a kid is such a big deal that I want to feel sure. I feel like if I'm going to make that kind of commitment, I want to feel more sure that it's going to be ok. And I don't feel sure.

If I thought we had a real chance, I'd have as many kids as you wanted."

Richard took Annasophia's hands.

"It's not that I don't want a family, and if I had a family I would obviously want it to be with you. But I would want them to have a real chance."

"But what does a real chance even mean?"

"I mean this. This life, the planet, everything.

If they could finally figure this stuff out like they've been saying they're on the verge of for the last fifteen years, then maybe I would feel better about having this kind of conversation. But right now, I just feel like having a child would be unfair to the child.

Don't you feel that way, knowing what we know? It's like they're going to be born with this punishment already, and they'll never be able to get out of that punishment. And *we* would be doing that to them."

"And we could also do beautiful things too. Like love them and teach them and prepare them to take on whatever happens.

I acknowledge that everything you're saying is true. But that doesn't mean we couldn't give a child a good life. Everybody's got their something. It's the other things. We can focus on the other things.

Aren't *we* happy?

We could make our child happy.

You're acting like Earth was some paradise before all this. But you know that humans lived in grueling conditions for over a hundred thousand years. I think you have to have some perspective. Compared to back then, our lives look magical."

"Yeah, but the difference is that they were moving forward, even if they didn't know it.

We know which direction we're moving."

"You're being dramatic."

"I'm not being dramatic." Richard let go of her hands and sat back against the wall. "I'm being realistic."

'You're being fatalistic."

"It's 2 in the morning; I'm being tired."

"Then go to bed."

"No, look, I don't want you to go to bed mad."

"Well, you're basically saying that you've made up your mind."

"No, that's not what I'm saying. I'm saying this is the first time we've ever talked about this. And I promise that it's not going to be the last. But I think we both need to think about it, and consider what the other said, and then talk about it again - preferably not three hours before I have to get up and go in to clean waste ducts for twelve hours."

"So you're not saying no?"

Richard stalled. He knew what she was doing. He might as well make her wait for it.

"I'm not saying no."

"So then it's a yes."

"Go to sleep."

"Ok, thanks for the yes, goodnight."

"Good night."

"Hey, wanna do it?"

"Good night."

Belonging somewhere doesn't make you any less lost

After so many meetings, so much rapport built, it should have been easier to get his mind off where he was. Still, the office, the building, all the buildings. The gray on the walls might as well have been a mirror.

He told himself. There was no reason to feel intimidated. Everyone else had been where he was, was where he was.

He could sense Galya wasn't disappointed, even if he was afraid to believe it.

Maybe she was dissecting him, in parts he couldn't recognize.

He suddenly felt the need to change the subject. To take the focus off himself.

"Can I ask you a question?"

"Of course."

"Why do you have these weekly meetings with us?

You have our logs and our reports. Why do you feel the need to…"

"To check in on you?"

"I guess so. Why do you make so much time? I would think you would have a lot more important things to do."

"I spent my life working on this technology. It only makes sense I would spend time making sure it reaches its potential."

Of course she would. Richard felt stupid for asking. It was presumptive and out of line. It was her program; she could do whatever she wanted.

He tried to recover. "You want to make sure the fuel you put in isn't contaminated."

"Sort of, but not the way I think you mean it. See, you think about this the same way everyone else used to think about it.

Whenever I would talk about this project early on, with the officials, with the funders, everyone thought about it in terms of travel. Like I was making a time machine. They were missing the point.

The potential of this technology is not the travel. It's what happens once a destination is reached.

I didn't figure out a new form of transportation. I figured out a new way to give birth.

Except instead of giving birth to humans, I've given birth to reality.

But what kind of reality depends on you. And that's why I take the time. The technology is done. It's your development that I have to focus on now."

"My development?"

"Yes, your development. This program offers you a potential unlike any before it. I want to make sure you reach that potential. Not just for all of us, but for you as well.

I recognize a part of myself in each one of you. But you especially."

She paused, as if she was deciding how much she should say.

"You know I used to work in a diner, long before I found myself in the clutches of science.

When I was fifteen my mother took ill, and I had to drop out of high school and go to work. I got a job in a diner downtown, where the tips were enough to support her and my little brother. I remember she had to write me a letter just so they would let me work that many hours. The manager said it was the law, and they were just looking out for me.

Anyway, at some point along the way, I read a quote attributed to Albert Einstein about how he often wondered how many times he had sat in a diner and been poured a cup of coffee by someone who might have been a far better physicist than he if they were simply allowed the circumstances to discover their potential.

I never knew if the quote was really Einstein's, but I knew the sentiment was true. Nearly everyone at the diner had a story like mine. Except most of them had larger families and worked at least one other place. They hardly had time to go to the grocery, much less read a book for pleasure.

I was somewhere in between. I actually knew I had a gift for numbers; I had seen it early on. But I came up in a poor school district and by the time I should have been accepting scholarships, I was busy rolling trays of silverware and scraping gum off the bottom of tables.

I used to wonder whether the downtown men who wore such nice suits and tipped me so well also had circumstances beyond their control that put them on their particular shelf. I used to wonder whether it was better to not know your potential or to know it and not be able to do anything about it.

No one had any reason to suspect I had anything to offer beyond a forced smile and a refill on their coffee. But I knew. I knew I was hiding something against my will. I was smothering it.

Then one day I saw an opening. Two men I'd never seen before came into the diner. One of the men kept making inappropriate comments, while the other apologized. I asked them what they did, and they said they were engineers for a government contractor, working on, as the inappropriate one put it, stuff that would be over my head.

I told him I wouldn't be so sure about that, and that I bet I was better at math than he was."

"You said that?"

"I did. And just like I knew he would, he said he would take that bet and then asked if I'd like to make it interesting.

So I told him his friend could pick a number between two and twenty, and that we would multiply that number by pi, and whoever got the most numbers correct wins.

If I won, they would get me a job where they worked. If he won, I would meet him after my shift.

Of course, if he won, I was going to inform him I was only seventeen and then threaten to call the police.

But I knew he wouldn't win. He was like me, in that he had no business playing the role he was playing."

"And that's how you got into this field?"

"That's how I acquired employment in this field. I would have belonged in this world no matter what. But belonging somewhere doesn't make you any less lost."

"That's an amazing story."

"Well, believe me, I didn't tell it to you to brag. It's quite the opposite. I know some people might listen to that story and focus on the choice I made. But those are people who want to feel like it's that simple. Those are people who want to feel ok about *not* seeing people like that little black girl pouring coffee in the diner.

I was there. And I know that the story isn't that I made my own destiny. It's that they walked in the door. It's that they sat in my section. It's that I was working second shift. It's that my boss didn't see me spending too long at their table and start yelling at me. It's a combination of a dozen things that one day."

Galya paused again, as if she knew exactly how much to say.

"I know when you came into this, you had it in your mind that this would be about you making choices. But it's not that easy. What you're doing is putting one thing in motion.

This is not about you *making* choices. It's about you understanding what choices are and how they are connected to the circumstances they are made within."

"But it is ultimately about choice. If you hadn't made that one choice at that one moment, none of this would have ever happened."

"You're right. Without that choice none of this would have happened. That choice was important. But only as important as every other choice I made thereafter. And considerably less important than the circumstances laid out before most of those choices."

"So which weigh more, the choices or the circumstances?"

"In the beginning, the circumstances weigh far more. In fact, for most people the circumstances are so heavy they feel like choices. But they're not. Real choice is not looking for the cracks in the wall. It's imagining how to create them.

I realized that there are certain choices you can make and if you make enough of those certain choices, the less circumstances there are to get in the way. Until eventually you're the one making the circumstances.

And, in time, you get to a position where you need to find someone else who can do the same thing. And that's what I believe I did with you.

I'm the only one who decides who gets in this program. I insisted that I have strict control over it."

Her words added so much weight, Richard's only option was to trust her.

"Why didn't you tell me that this was your technology in the interview?"

"Because that moment needed to be about you, and not about me.

I know that you've been discouraged lately. But you need to know that I'm not. I believe in you. You're going to find the place. We chose you for a reason. I chose you for a reason.

You see every visit that doesn't work as a failure. I see it as your education. I see it as you developing. Without this education, without this development, you can't make the right next choice.

Finding my way here to this moment wasn't about one choice; it was about each next choice I made. Just like you finding the place is not going to be that chance you took answering the queue and coming in for the tests. It's what you're going to take a chance on tomorrow and the next day, until you find one set of circumstances that appear too great to overcome. And then you'll find a way to make it happen.

That is what's going to get you to the place. I believe that."

"I wished I believed in myself the way you believed in yourself."

"That's because you weren't there. I know, the way I tell it, it makes me sound quite bold. But the truth is I was full of insecurities. I often cringe when I think of how many opportunities before that I let pass because I assumed it was pointless to try.

It took me five times walking past their table before I mustered the courage to start that conversation."

"What finally made you do it?"

"I'm not really sure. I guess I just couldn't imagine my whole life being just out of reach.

I've done all I can do to get us to find a technology that our circumstances didn't allow for. I believe that technology exists in another set of circumstances. I believe it's there.

But you have to believe it's there. And you have to continue believing that."

"I have one last question.

How many of the others have you had this same conversation with?"

Galya smiled, slightly. "And it's demonstrations like that that absolutely do make me ask myself why I make the time.

After we find the place, maybe we can work on developing your sense of humor."

Someone to feel connected to

It was a lull. That space between one discussion and another. They were all comfortable enough in the room that it would be no surprise for any of them to start a new one. Within this landscape, of chairs and walls, their familiarity had become a necessity. They might as well have been naked.

But they weren't. There were still things they had not gotten to.

They waited.

Rachel ended the lull.

"Am I the only one that feels like I'm sexually harassing my partner?"

"What do you mean?" Normally so monotone, Abbott sounded slightly alarmed.

Ming Yan, however, was not alarmed. "I know exactly what she means. We're gone for weeks sometimes. And when we come home we're so excited to see them, we can't wait to get it on. Yet they can't understand, because for them, they just saw us a few hours ago."

Even Clayton could sympathize. "Yeah, they don't tell you in orientation you're going to end up feeling like a lecher."

"My partner thinks I'm a sex addict. And of course I don't expect them to be in the mood every day. But then I also know if we're not together, it may be another two weeks before I even have the chance."

"And that's no guarantee either." Richard had never felt closer to Rachel than in this moment. He had been waiting for someone else to bring this up.

Ming Yan added, "Factor in periods, and I might not get laid for the equivalent of months."

"She thinks I'm weird because this is the most romantic I've ever been, and I can't tell her I'm acting this way out of desperation that she might not sleep with me that night." Rachel shared this almost like a confession.

Clayton as well. "Yeah, it's bad enough we can't share what we see when we're out there, but to not be able to let them know why we have the needs we do is beyond frustrating. It makes me feel things I don't want to feel about my partner."

Then, as if to start a war, Bryant just shook his head and laughed.

Rachel came to Clayton's rescue, and her own. "What's funny about that?"

"What's funny is listening to you whining because you're too old-fashioned to take care of your needs.

I wish I could sympathize, but nobody's forcing you to be monogamous."

Clayton countered. He wasn't the slightest bit self-conscious, even if he did feel the old-fashioned remark was aimed at him.

"We made a promise."

"I made a promise too. But just like you, I made a promise under a set of conditions that don't exist anymore."

Norman faithfully chimed in. "He's right. If we're not allowed to share what those new conditions are, then I don't see how the same rules apply."

"They apply because the vow I made was for better or worse." It was as if Clayton's wife was in the room with them.

"And you can keep those vows in this world. But out there, no way." There was a smugness to Bryant's tone.

"Why should any of us have to live up to the same expectations as our partner, when they're not out there?"

"Is that what you tell yourself?" The distaste in Rachel's voice was unmistakable.

"You're damn right it's what I tell myself. For what I have to go through and the good I'm doing, I think I deserve a little release."

"And I bet you don't give a second thought to the consequences, do you?"

Bryant leaned in. "And what consequences would those be?"

"I don't know, maybe how many kids you've left behind. You're out there doing your thing with who knows what, and have you ever stopped to think - once this is over, how many kids in however many different realities you'll have left to grow up not knowing who their father is?"

"I don't have to wonder. I know.

I know because I keep track when I go forward on the branch."

"You what?" Richard almost surprised himself; he normally avoided confrontation in Group. It sounded as if he was almost ready to fight.

"Don't look at me like that. If we can't have kids here, while we're in the Program, then why not at least be able to see what your offspring would look like? And before you get all self-righteous, that's the reason why I always request extra diamonds to bring. Those kids are taken care of. And the women lucky enough to not get knocked up? For them, it's like sleeping with me is hitting the lottery."

"You're disgusting." This wasn't the first time Rachel had said those words, specifically to Bryant.

"Oh, yeah? Am I disgusting? Or are you just jealous that I found the loophole in the Agreement, and the loophole can only be taken advantage of by the men?"

"Screw you!"

"Hey, I didn't write the Agreement. I just read it. You were given the same justifications I was, and you signed on just like I did. It says, 'Until we find the place, neither participants nor partners are allowed to bear children.'

But guess what? We don't know if we're ever going to find the place. Hell, we don't know if we're going to even come back next time we go out.

I've visited my kids. I've spent time with my kids. Do they know it's me? No. But I know. And you can think it's unethical all you want. But at the end of each long long day, I haven't broken any more of my Agreement than you have of yours."

"Is that all you care about? Is that all you see yourself as? The Agreement?"

"Yes, and that's all you are too. Once you signed that Agreement, that's what you became.

And that Agreement has granted us all a lot of license. Things we wouldn't normally be able to get away with.

You think you're better than me because you keep it in your pants? I know what you've done out there. Even if you don't say it here in Group, I know what you've done.

Because we've all done it.

So don't judge me too close. Or you might realize you're judging yourself.

This isn't about sex. It's about intimacy. It's about you living in two different worlds. And one world doesn't have the same rules. And when you come home to this one, you want someone to feel connected to.

But it's not like that anymore.

So you can think I'm disgusting all you want, and you can keep your vows to your partner all you want. But at the end of the day, I'm the one you really come home to. I'm the one you can share with. I'm the one that's there for you.

We're the only ones who know what you're going through, that understand - any of this."

It's not you

It was still strange, to be seeing her like this. Like seeing someone's baby pictures. Except she wasn't a baby. She was a young woman, on the verge of finding a way.

Before, decades later for her, Richard had only ever seen Galya in her office. He thought about how this place must have looked to her, and how all the things that might have haunted her senses during her shift were now objects of nostalgia for him. Not that he had grown up going to diners like this one. But he had seen pictures and movies; he had heard people talk about them. Such fondness.

Galya was only seventeen, and she looked so worn out.

Richard knew her break was getting ready to end. She could only sit in the booth with him for so long before the manager noticed again through the order window and said something.

There was time though. He could sense it was coming. The disbelief.

She only needed skim the folder.

"This can't be real."

"I understand this is a lot to process, but I can assure you it is real."

"Why would you think I would believe any of this?"

"Because I'm giving you this."

Richard slid her a piece of paper with two sets of numbers written on it.

"These are the winning numbers for tomorrow night's lottery drawing. That is for your assurance.

And these are the winning numbers for June 27th of next year, the largest unwon jackpot to occur over the next decade. That's all the money you should need to put into motion whatever it is you're going to put into motion."

"So you don't know what I'm going to do?"

"I don't know what you're going to do this time."

"What does that mean?"

"You're assuming this is the first time I've had this conversation with you."

"I don't get it."

"You've bought tomorrow's ticket three times, making the other winning numbers worthless.

You've bought the June 27th ticket twice and simply retired young and rich.

But mostly, you've done the right thing and attempted something and it just didn't work out. Not to mention, there's been a few times, you got up and left before I could even get this far."

"So why do you keep trying? Why me?"

Richard gently brought the folder to the middle of the table, turned to the last section of pages, then eased it back beneath her hands.

"Everything you've attempted and every obstacle you've encountered, when you did attempt it, I've detailed in here so you can avoid your mistakes. With the drawing so many months away, it should give you enough time to digest the information."

"You didn't answer my question."

"I didn't answer your question because I shouldn't have to. There's technology in what I'm giving you that your world would not realize for decades, and I'm giving it to you."

"You think one person could take the technology that failed you and make a world so different it would produce technology that could save you? You really think one person could make that big a difference?"

"First off, it's not that the technology failed us; we simply acquired what we have at too late a stage.

Second, I'm empowering you with the collective work of millions. And I'm empowering you with enough money to hire an army of brilliant minds to implement it.

It's not about you. It's about what could be done, and what I believe you could make sure happens, with these tools.

I used to think it was about one person or one event. I used to think that humanity would just find its way if it could hold onto that hero for a little bit longer, or I could prevent that bad thing from happening.

But the truth is, you're not our only hope, and I'm not the only one working on this. There's an infinite number of scenarios where this works out for us, and there's an infinite number of scenarios where it doesn't.

Unfortunately, I only have a *finite* number of visits I can make. So if you can't grasp why I might keep having this conversation with you, then I'm eventually going to quit having this conversation with you.

It's not you. And yet I'm here, still, offering you these tools."

Richard pressed the tips of his fingers down onto the pages.

"If you really need to know *why you*, read through the last section. There's a bio with your original future inside. It's pretty impressive."

Galya looked at the folder and the data drives. She contemplated what she might learn. The details of her failures, and the one success she could only know from notes on a page.

She abruptly grabbed the folder from underneath Richard's hand and gathered his cup and silverware from the table.

"I have to get back to work."

"Ok, but listen. Whatever you do, check the numbers tomorrow. This isn't a joke. I chose you because you chose me."

"How did *I* choose you?"

"Read it. I promise it will make sense."

"I have to go."

Come with me

He'd been standing there for nineteen minutes, including the minute since the scan downstairs had signaled arrival. He only looked at the door. There was nothing else. Not the furniture, not the decorations. Not the colors, not the attractive gray. Not the clean, not the secure.

This time he was really looking.

The knock came, and he opened the door. Annasophia walked through it.

"Hi." She tried to smile.

"Hi." He was too nervous to smile.

"Sorry I'm late. Traffic was bad, and I was watching my nieces until my sister got home. I don't remember leaving any of my stuff. What is it that you found? I know I--"

"I'm leaving here, and I want you to come with me."

"What? What are you talking about?"

"I'm leaving the Program, and I want you to go away with me."

Annasophia shook her head. "You know you don't have to do this. You know what? I'm not going to let you do this. I know how you feel about your job, and I'm not going to be the one to take that away from…"

"You were right. You were right and I was wrong, and I've been wrong this whole time and you're the only person I should have been listening to and I didn't. I didn't trust you. I didn't trust you, and I'm sorry. And I can't take that back, but I can try and do the right thing right now. And that's what I want to do. But in order for me to do that, I need to know if you trust me.

Do you trust me?"

"I don't know. I really don't. It's not like some switch you get to flip on where you just say you've changed and then it's all better. Instead of asking me if I trust you, you should be showing me what you've done to earn my trust."

"That's why I asked you here."

"No, that's not why you asked me here. You asked me here because I didn't come back when you thought I would, *how* you thought I would. And now you think the way to get me back is by quitting your job. And I don't want that. I don't want that to be my decision.

Do you even see how unfair that is? It's like you're asking me for permission to do the right thing."

"I'm not asking permission. I'm asking if you'll go with me. The decision for me to leave the Program has already been made."

"What are you talking about? Did you get fired?"

"It doesn't matter. All that matters is if I can convince you right now that everything has changed, that everything is different. I know this is out of nowhere and I know that I'm asking a lot of you. But it's because I don't have any other choice.

All I can do right now is ask you to be with me."

"Look, Richard, I'm glad that you're quitting. You know how I feel about your job. You know what I think it's done to you. But just because you're burned out, and because you're burned out - *now* you're ok with coming back to the Stretch, it's not the same as…"

"I'm not coming back to the Stretch. I'm not coming back to anything.

You remember when we almost moved to the Northern Headlands, and you were willing to leave your family and friends behind so we could try and make a life for ourselves?"

"I remember that we were young and dumb."

"I remember us regretting not going."

"What are you asking me?"

"I'm asking if you remember who we were. I'm asking if you still love the person you almost went away with. I'm asking if you would still leave your family and friends, and everything you have here, to be with that person.

Because I'm going to be that person. Except I'm going to be even better. And I'm telling you that I can give you everything you want; I'm *going* to give you everything you want. Everything we've ever talked about, I'm going to give to you. And I promise you that we'll have a baby. No more blaming our situation; no more rules that I'm not allowed to explain; no more telling you it's right around the corner. No more waiting. We can finally do it. And we can do it now.

But in order for that to happen I have to get you to make a choice, right now. And I have to know that you trust me."

"Are you in trouble? Because if you--"

"I'm not in trouble. I just don't have a lot of time. I promise, as soon as I get a chance, I will tell you everything and all of this will make sense."

"But I'm not asking you to do that. Seriously, Richard, I don't want to hear about your job. I know what you do is secret, and I don't want you to do anything or tell me anything that's going to get you in trouble or get me in trouble for that matter. The less I know about your job, the…"

"I'm not talking about the job. It's not about the job. The job is over. The Program is done. It's all over.

This is not about anything but me and you.

I know that the way I'm doing this right now is crazy, but I'm trying my best to do the right thing right now and I don't know any other way to do it."

"What do you mean 'do the right thing'?"

"I can't tell you. I can only show you."

"Look, I can appreciate that whatever it is you're not telling me is probably for my own good. But asking me to trust you, in hopes that you're going to prove why you deserve it later, is totally unfair. Never mind putting me on the spot like this and asking me to decide something I don't know what I'm deciding. None of this is ok."

"You're right. It's not ok. But it's the only way this can be."

"And you really think you can expect that of me?"

"I know I can't. But it's what I'm asking."

"If you know how unfair it is and you know you can't expect it of me, then you also can't be let down when I say no."

"Please, Annasophia. I'm not asking you to forget what I did. I'm not even asking you to forgive me right now. I know I have to make up for all that. I'm just asking. Are we worth taking a chance?"

"If you were asking me that, I would say yes. But you're not.

I'd like to leave now."

"Please, just one more minute. Just take one minute. I won't say anything. Just stand with me here for one minute."

Annasophia relented. She knew it was unfair. She knew what he was doing.

Still, it was something she could give him.

Neither of them spoke.

A little over fifteen seconds in, they both started to cry.

They cried for another twenty seconds, before Annasophia took a deep breath.

"I can't do this. I have to go."

She walked past Richard, opened the door, and walked through it.

Richard didn't call out. He didn't begin begging. He just swallowed and calmly requested, "Before you go, can I just ask you one thing?"

Recognizing the resignation in his voice, Annasophia turned her body. "What?"

"When you walked in the door just now, what could I have said or done to convince you?"

won't pop and somehow I'll be able to just catch it and hold it in my hand. But that doesn't happen.

And here I am, once again, in a therapy session with all the other children, talking about the bubbles we chased."

"Except you're not children. You're courageous men and women doing a very important job. And, as for your dream, you need to remember you're not chasing the bubbles. You're blowing them. Chasing them implies someone else is making the choices for you. But you're making the choices. You're making the bubbles."

"And I guess that just makes the dream worse."

"How do you mean?"

"How do you mean?!" Ming Yan didn't expect sympathy from Abbott. But for him to feign ignorance?

"Have you not been sitting in here with us for months now? Have you not been listening? Do you grasp the consequences of what I had to do to make those bubbles? No, you don't. And, you know, I probably don't even grasp the consequences. And I was there."

"You choose your methods. We've given you autonomy. If you regret your decisions, that's not our fault."

"Not your fault? Why do you think we're here? It's somehow my fault that I'm out here making impossible decisions to save us from a world that I didn't create?"

"You need to calm down." Abbott was stern, while making sure not to raise his voice.

"No, you need to get out of my face and quit talking to me like you understand anything I'm going through. Talking to us like you're our friend or something.

Why are you even here? Oh right, you're our manager. I know you call yourself a facilitator, but isn't that what you really are, a manager? And we're just workers, digging in the dirt, trying to find some buried treasure that you'll end up controlling and we'll never get credit for."

Ming Yan looked around the room. She could see the sober looks of gratitude on her friends' faces.

Shaking her head, she proceeded, "You know I thought this all sounded great when I started. Being part of this big breakthrough. Getting to go back and actually see the past. Getting to live out adventures better than the movies, and getting to live out a new one every day.

And now here I am, coming home each night, hoping it doesn't show that I've aged two months. Each night. I'm aging literally overnight.

I feel like I'm living a lie because I can't tell my partner anything about my life, about who I am now, who I've become. What do I say when my wife asks me, 'How was work?' I'm just supposed to look her in the eye and lie to her? Every night I get to tell a lie to the only person I've ever loved, who's ever really loved me.

When we first came to Group, you said we would adjust. But I haven't adjusted. Because you can't adjust to this. Because this isn't normal."

"I know this isn't normal. And you have every right to feel the way you do. But you have to keep telling yourself that this is only temporary."

"No, you have to keep telling us that. Because that's what you're supposed to do as a manager."

"I'm not your manager."

"Then what are you? Because you're not facilitating anything. And you're definitely not my friend.

You talk to us in this calm voice and act all patronizing, but you know what you really are? I'll tell you what you are. You're a prison guard.

That's right. You're a prison guard, and you don't even know it. And we're in prison, and we don't even know it.

We're not heroes; we're prisoners. And Galya is the warden.

I bet Galya gets to leave the Zones. She does, doesn't she? Does Galya know what it feels like to be jealous of your own partner because *they* have the freedom to leave the Zones, to see their family and friends outside? Is Galya surveilled 24/7 to make sure she doesn't say anything about the Program? Does Galya understand what it means to do this job?

It's bad enough I signed a contract to live in a prison and treat my time with my partner like some kind of conjugal visit. But to try and hold onto my conscience while I'm living under a different set of rules, a different set of laws? That's what prisoners are asked to do. And then no one understands why the prisoners act the way they do."

"Is that what you want? You want to be free? You want to live out there?

You know how nice it is in the Zones.

It's going to keep getting worse out there. Pretty soon there's not going to be seasons anymore. You romanticize this idea of freedom. But if I let you out there, in one week you'd be begging to get back in, to get back into the Program. You *know* that there is *nothing* out there for you.

In here, you have something. In here, you matter. You get to be heroes. You get to become saviors. You get to live things and see things that no one else on Earth will ever even fathom, and you…"

"Do you know how many people I've killed?" It was the first time Ming Yan had screamed in Group since hearing the news of their third death. It was the first time she had ever screamed at anyone in the room.

"I'm not talking about the monsters. I'm talking about…

I thought I was signing up to do something good. But we're not heroes. At best we're pirates, looking to steal someone else's technology.

Do you know how many people I've killed?"

"And every person's life you had to sacrifice will keep millions more safe. This isn't some thought experiment in a college ethics course. We don't get to live in the ivory tower; we don't get to rest our faith in our advanced degrees. We don't get to be self-righteous.

This is real, and you are here to do the best that can be done within the parameters of your mission. But you have to remember: Every terrible thing you find yourself doing is in service to a better world. You can't focus on the lives you've taken. You have to focus on the lives you're saving."

"Is that what you think we're still doing? Saving other people's lives? We're doing good to fumble through this and make sure *we* stay alive.

It would be different if what we were looking for had a face. Something we could recognize. But instead we're supposed to make it up. Except at some point I started losing my imagination. Until eventually it was all used up. And now I just feel like I'm going through the motions. Because this feels futile, and I don't really believe in anything."

"I understand what you're saying, and I want you..."

"Do you? Do you understand?

Because when we lost Devin and Manriquez that third week, you didn't halt the Program and figure out some better technology to keep us safer. You didn't do anything like that. You just kept interviewing. You tweaked the criteria and kept interviewing.

Their families don't even know what happened to them. We don't even know what happened to them.

I used to be afraid someone else would find the place first and I would feel like all my work had gone to waste, that my life had gone to waste. And now I hope one of you will be better than me, just so it will be over.

I'm tired of playing God."

"You volunteered for this."

"Did you hear that? We volunteered to play God.

Except I didn't volunteer. I checked my opportunity queue."

Suddenly Ming Yan felt alone. She looked around the room. The gratitude was gone. Now they just looked like they were scared, of what she might say next.

Ming Yan gave them what they were afraid of.

"What? You think they sent that out randomly?

We fit a profile. Our backgrounds. The things we've done in the past. What our files say.

You know the surveillance they have. They selected us before we ever got the interview. We got played, and we're still getting played.

We're not heroes. We just fit a profile. Every reason they gave us for why they chose us was flattering. At first, I thought it was just to build and maintain our confidence. But now I think it's because they know it's the only way we'll keep doing this. It's so they can keep playing us.

Why do they keep telling us how autonomous we are? Why do they keep telling us how principled we are? Why do they keep telling us how important we are, how special we are?

They tell us that so we won't question.

Why do you think we're on a need-to-know basis? It's so we won't know what else is really going on here. And then they tell us we're autonomous."

"You are autonomous." Abbott was still calm.

"We're autonomous as long as we follow orders."

"You're autonomous to work within the terms of the Agreement."

"That's just a fancy way of making us think that the orders were our choices."

"And when we find the place, are you going to say that these were just orders? Or are you going to take ownership of the autonomy you were given and give credit to yourself, to the choices you made?"

"That's assuming we're getting closer. But we're not. We're not getting any closer. We're not.

They told us this operation's been going on for a few months. What if it's been going on for years? Why did they stop with Richard? Why did they stop at seven of us, excuse me - ten of us?

If it's that important, why only us?

What if there are other teams doing this and we don't even know? What if there are dozens of other teams, doing the same thing, and none of us can find the place?

How do we know they're not lying to us? We don't know anything. How do we know they're going to live up to their part of the Agreement? How do we even know we're going to get out of here once we find the place?

They can't be sure we won't tell someone later. Are we going to have to stay in the Zones forever, even after you've employed the technology?"

"Would it matter? Even if everything you're saying was true, which it's not. But even if it were true, it doesn't change the fact that outside there are storms coming that we will never recover from.

What you're feeling, what you're going through. It's understandable and it's regrettable. And if I could change it, I would. But I can't. Because this is our only shot. What's going on out there isn't going to stop until we stop it.

You had a choice. And you made a decision of your own free will. We didn't force you to sign the Agreement.

You signed the Agreement. You did this. I'm just trying to help you get through it."

"You keep talking about choice, but you yourself said it. What other choice is there for people like me? Go back outside the Zones and hope I make it past my 40s? There are no choices anymore. They all got made. And you know who made them? You. The same people that are managing the Program are the same people that made all this happen.

I didn't do that. You all did that.

And yet I'm expected to rationalize my sins so you can somehow be redeemed of yours.

And yet I still feel like I have to do this. I feel like my sins are chasing me and I have to keep running. Even though I know it's not my sins I'm running from. It's yours.

Why am I running from your sins? Why am I running from the sins of men who will never have to admit to what they've done, who will never even believe they could have sinned?

These are your sins. I'm tired of running from your sins.

I'm tired of…"

"Stop it!"

Ming Yan was taken aback, as was everyone else in the room. It was the first time they'd heard Abbott raise his voice, much less look out of control.

"Stop all this!

I've tried to be considerate and accommodating. But this is where it ends.

You think being philosophical is going to change what it looks like out there? You think this navel gazing is going to make you feel better for your failure at being a hero?

You're right. You do have orders, and if you don't want to obey your orders, if you don't want to fulfill the terms of the Agreement you signed, then we will have no other choice… than to fully execute the terms of that Agreement.

So you had better think, tonight, about whether you want to avail of the opportunity you've been given. Because you're right. You haven't found the place, and you may not be any closer to finding it. So if you're too tired to keep trying, or to show us that you're serious about this, then I will show you how right you are.

We're not friends. And it would be no more inconvenience to me than having to inform Galya we need to conduct another interview."

A morbid curiosity

As they sat talking, finishing their lunch, a guard would occasionally walk by and do a double take. Thinking it was the same Galya they knew.

But her hair was different, as was her style of clothing.

Even for Richard, it was strange, to see her like this. Like seeing someone's identical twin. Except she wasn't a twin. She was the same woman. She had just chosen a different route to get there.

On their walk earlier, she described what was on the data drive she had brought, rifling through the science and how it could be applied.

The technology. It existed. She had proven them right. She had spent her whole life proving them right.

When she thanked him for not giving up on her, he couldn't help but worry how this world must look, to her. Things she had only seen in pictures and words. Such failure. He feared it would haunt her when she returned.

The only thing that consoled him was the relief. To finally be sure, of anything. He finally felt ok, with his hope.

There was time now. He could sense it coming. Salvation.

"I see you're wearing the watch we gave you."

Richard smiled. "Are you kidding? This is my trophy. All my souvenirs from before were from failed visits.

Plus, I love it. The functionality is so far beyond what any of our devices can do.

It's really something. Thank you."

"Well, thank you for the tour. I appreciate your people allowing me to see all this."

"I think it's a modest request considering what you're offering us.

I'll tell you though, it's a bit bizarre giving a tour of what it's like right now, knowing that everything is about to change.

How do you feel now that you're here?"

"I'm not sure really. I felt almost guilty wanting to come here. I know it's a morbid curiosity, but I've been leading a group of scientists to make models for a climate we've never seen and frankly have had a hard time imagining outside of those simulations. They only believe it exists on faith that what I'm telling them is true and because the science is theoretically possible."

"How does what you imagined measure up with what little I've shown you?"

"Let's just say I understand the level of urgency."

"Try watching it get worse.

It's like I told you earlier, I've been part of a group working on this. And at first, we made it a competition with one another. It made the job fun. But as it got worse, especially recently, we found ourselves more sober about what was at stake, and it really became more of a race with what was coming."

"Did it still feel that way when the others found out that you had done it?

I don't imagine everyone fully shook that spirit of competition."

"Honestly, I think they were just as happy as I was. We've shared so much with each other along the way, it really did feel like a team winning rather than just me."

"But you must feel some sense of victory in being the one to solve the maze."

Richard smiled. "I might have done a little dance in front of the others. Maybe a couple dances. But mostly it was just a sense of relief."

"I'm glad it worked out the way it did.

It's strange having waited to see you for all these years, and for you this was a few days ago."

"Three days to be exact."

"Is it strange having me here?"

"You mean do I keep expecting everything to turn inside out because we've broken some sacred science fiction rule of interdimensional travel?

Of course it's strange. But, for me, it was actually stranger to see you when you were younger. I'm used to you being this age."

"Can I ask you something?"

"Sure."

"Do you normally go back and visit people you've affected?"

"I never do. The only reason why I came back to see you is because you already knew everything, and I figured if it was ever going to work you would be the one leading the way. Just like here."

"Is that the only reason?"

"I don't know, I guess I felt like I owed you."

"In what way?"

"Because you're the one who put me here. I know it's not really you, but it still feels like you're the same person. That probably sounds stupid."

"Not at all. I completely understand. One of the things that convinced me to take the direction I did was being aware of my life's impact here."

"So you do feel like what you did here is yours also?"

"I'm not entirely sure. I know I have no more ownership over her actions than I do anyone else's, or everyone else's for that matter. Yet I still feel this weight.

I'm not sure whether it's the weight of knowing what I was capable of and knowing what was up ahead - and how that left me with a feeling of responsibility to act. Or if that weight was the accumulation of all our actions up to this point, and it was more a question of how to help carry that weight forward."

She stopped. And for a moment, she looked as if she were embarrassed. Like she was rambling.

It amused Richard. He had never seen Galya utter a word that sounded anything less than measured.

It was nice.

She elegantly recovered, "Why did you decide to let me know who I was here?" It was something she had been wanting to ask her whole life.

"I'd love to tell you it was wisdom beyond my years, but really it was trial and error.

I actually didn't tell you the first few times. That was one of the things I'd started doing differently. I just figured I would try it and see what happened."

"Why didn't you tell me from the start?"

"I was afraid that's where your focus would gravitate. That it might serve as a distraction."

"That's a fair concern. Is that why you didn't leave me the travel technology in the folder?"

"I didn't have that information. I don't even understand the technology. My position is only privy to the coordinates. They keep the actual technology as much a secret from us as they do the public."

"Is that why you said you're not allowed to leave this area?"

"Yeah. At least for now."

"I guess that makes sense. Though it's funny that you're the only reality they would have to keep it from."

"Why do you say that?"

"At first, I had assumed your not including it was a defense measure, guarding against the possibility that the possession of this technology by any other realities could potentially interfere with your progress.

Of course, then I realized it wouldn't make a difference, and that even if someone found their way to meddling in your timeline, it would merely result in the creation of another new reality."

"Leaving this one still intact."

"Correct. The only lasting change that can be made to this reality has to be made in the present."

"And that's what makes the secret only valuable here."

"Correct."

"Now that you mention that, maybe I'll see whether I can let you take back the recordings I was telling you about. Like you said, it won't really matter for us who hears it where you are."

"Please do. I would love that. I was intrigued when you mentioned them earlier."

"I have a ton of them. The logs we're required to keep are normally more technical, and I just couldn't stand that being all there was to our history."

"I understand. From what you described, these recordings sound fascinating. And you're absolutely right. Such a moment deserves to be experienced and not just told about."

"Aside from their importance, I just think it would be nice for somebody to hear them."

"If you can't be famous here, you might as well be famous there."

"Yeah, maybe so."

Richard grinned, as if already contemplating his fame.

To which, he posed, "So why didn't you go for the fame?

Even though you did succeed, you had no reason to believe you would. Why didn't you go after what you already knew you had succeeded at? Just the magnitude of such a discovery would have made you a science god."

"That's assuming *we* would have let anyone know we had this kind of power."

"Yeah, but even if only a few people knew, wouldn't that have been worth it?"

"For a moment, perhaps. But you had already shown me what the far greater challenge was, and actually what the greater reward was."

"I get it. I guess I'm just surprised it wasn't more tempting. Knowing that you had this technology in you."

"It wasn't in me. It was in my pathway.

You gave me the ability to try down a different path, one that wouldn't be conceivable without your visit.

And the more I thought about it, the more I realized I wouldn't be able to let it go."

"Why spend your life trying to figure out what, at least in this world, you had already figured out?"

"I thought it better to let others pick up that path."

"Knowing that it was possible. Makes it more a matter of when than if."

"Correct."

"Not as much fun trying to discover something if you know, for certain, that it can be discovered."

"Correct. Except using the word fun implies such discoveries are always a prize, something which becomes a matter of pride. In this case, in particular, such knowledge wasn't a matter of pride."

"If it wasn't a matter of pride, what was it?"

"I'm not sure what you would call it. I just know there was no comfort in exploring the problems that were left for me to solve."

She stopped again. It was a moment to be sad.

Richard waited a few more seconds.

"We should probably get going here in a bit. I know we've been on the move the whole morning.

If you'd like, we can go back and you can take a nap, and in the evening I can have someone take you over to the dinner."

"That would be nice.

I have to say, I'm a bit nervous about tonight."

"You shouldn't be."

"What's she like?"

"She's like you. I mean you both look the same. You're both brilliant."

"Do you see any differences?"

"It's hard to say. Aside from little things, you seem like the exact same person. Then again, I don't know whether I know you well enough to even say."

"Me or both of us?"

"Now that I think about it, I guess both of you."

"What did she say when I requested to meet her?"

"When I showed her the progress you helped your people make, she was thrilled at the idea."

"I hope it isn't too awkward."

"Each of you talking to yourself?"

"Yes."

"Hoping everything won't turn inside out?"

"Perhaps." She smiled thinking about the prospect.

"I think it's going to be cool. Getting to find out what interesting things happened in your respective lives, how you reacted to different situations. Seeing what could have been. I think you're going to have a lot to talk about."

"We'll see."

"Do you mind me asking why you wanted to meet her? Aside from normal human curiosity and it being the most bizarre awesome experience anyone could ever have. Aside from that."

"Aside from that, I think I want to see how much of myself I recognize in her."

"I wonder if you'll be finishing each other's sentences."

She smiled thinking about the prospect. "Perhaps."

I'm here because I killed God

The lamp came on, as Richard started to come to. When he heard the click, he thought it must be Annasophia. But why would she have come in the middle of the night?

When he could finally make out light from shadow, Richard saw her, across from him, sitting on the chair he usually laid his jacket on, a small gun resting in her lap. It wasn't Annasophia, but it also wasn't a stranger.

Her hair was different, as was her style of clothing. But she was older than seventeen, if only a few years. But he wouldn't know any of that anyway.

Her face might as well have been a random baby picture, of an acquaintance.

He checked to see if he had been restrained. He hadn't.

He sat still, and stared.

"Hello, Richard." She knew his name.

"How did you get in here?"

"You call them coordinates."

"Who are you?"

"I'm someone you did a favor for, and I'm here to return that favor."

"I don't know who you are."

"You don't know who I am because you haven't met me yet."

"What are you talking about? Who are you? How do I know this isn't a trick?"

"Because you know that this is the most secure building in the city. And because I can tell you that you were just dreaming about sinking on a boat."

Richard looked around the room. For what, he wasn't sure. Something to trigger an explanation, or at minimum to provide better bearings.

"What do you want from me?"

"I'm here to recruit you."

"Recruit me? For what?" He was sure she couldn't be much more than 20.

"I'm here to recruit you to be the person you wanted to be, the hero you tried to be."

"You're here to recruit me with a gun?"

"The gun just buys me this conversation."

"And what does the conversation buy you?"

"I'm hoping it buys me your trust."

She placed the gun onto the dresser next to her, the barrel pointed away from Richard.

"Now, in a minute, I'm going to prove to you why I'm here. But first I need to prepare you for it.

Richard, have you ever really asked yourself what it looks like if you're successful with this *Program*? You've been earnest in your search for this one thing, even though you have no idea what that thing looks like.

Did you ever stop to think that maybe your faith in your intentions has you under the illusion that those intentions are shared?"

"I don't understand what you're talking about." How could she know about the Program? No one on the outside knew.

"You don't understand what I'm talking about because your focus on winning the little game you've been put into has kept you from asking certain questions."

"So what am I not asking?"

"Why are you doing what you're doing here?"

"To save my people."

"You say that as if it were a simple thing."

"What's not simple about wanting to save your people?"

"It depends on what you mean by save and what you mean by your people."

"Why don't you just come out and say what you mean? Why all the riddles?"

"I'm sorry. I really am. I still haven't figured out how to convince you of something I know you're not ready to see."

She stopped, and let build. What she was about to say.

"In five days, you're going to set out on a course that will kill tens of thousands of people, your people."

"That's impossible."

"It's true."

"How? How am I supposed to do this?"

"By visiting me."

"You're lying."

"I wish I didn't have to be here. But it's more than just what we wish for."

There were plenty of reasons to not believe her, to not listen.

"Are you here to prevent what I did, what you said I'm going to do?"

"No, what you did was a good thing."

"A good thing?"

"You know there's more to the creation story than what we were told.

I don't mean that it didn't happen. I mean that it's not the only version of that story that happened; it's just the one we know."

She then rose from the chair and began to slowly pace the room as she spoke, leaving the gun within a leap's reach on the dresser.

"What if Eve had eaten the apple, given it to Adam, God punished them and threw them out of the Garden, and then Adam in a rage killed Eve? That's not inconceivable.

Who's to say that very scenario didn't actually happen, and God simply started over?

And the next time, God punished them, leaving Eve to feel this was unfair, for which she decided to run away. What if that happened too, and God just started over again? Except this time, instead of Cain killing Abel, he couldn't bear the shame that it would bring his parents. So instead he killed everyone, including himself. What if that happened?

What if all of those happened, long before the story we know happened? Except instead of just starting over, God allowed all those scenarios, and a lot more, to play out in their own way. Maybe he saw it as a kind of game, setting each story in motion just to compare their outcomes. And then, applying his own arbitrary criteria, he decided after a certain amount of time had gone by - to pick a winner. Leaving all the rest to perish.

The point is not whether we're the winner or whether we're going to perish. It's that all this time we've been in a contest we didn't know we were in.

The original sin wasn't eating the apple. It was God setting us in motion within these parameters. We've been carrying our creator's burden ever since."

"Just tell me why you're here."

"I'm here because I killed God, and I want you to help me go back and relieve all the other creations from their burdens."

"And why do you want me?"

"Because you gave me the chance to make right what I did, and I'm offering you the same chance.

I owe you that."

"What did I do wrong?"

It was enough. She hoped.

Easing onto the foot of the bed, she took the heart from her watch and set it down between them.

"Play recording."

Still capable of hope

Richard was setting the food on the table. Annasophia started to speak, then stopped. The train was coming. Even on the 18th floor, you could hear everything, outside.

Annasophia noticed Richard was acting strange, almost catching himself in a smile.

She had also noticed him looking around the room, in different rooms. Looking at the chips in the paint, the walls scarred by who knows how many couples, how many families.

Richard had no idea how much he would one day miss those chips, those scars. He had no idea how much he would regret that smile.

They had been living there long enough that he couldn't really see anything, but getting out.

The noise from the tracks finally faded. Annasophia spoke.

"You look like you're dying to tell me something."

"Ah, come on. I thought I was doing a good job hiding it."

"Worst poker face ever."

"It's not that bad."

Richard tried to make a straight face, then broke character immediately. They both laughed.

"Anyway, I was going to wait until I found out more, but I guess I'll go ahead and tell you.

I might have a possible interview for a job."

"Yeah?"

"Yep. And it's in the Zones."

"The Zones! What would you be doing in the Zones?"

"I'm not even sure yet. I just know it has something to do with one of the geo-engineering programs."

"What do you know about geo-engineering?"

"I don't know anything about geo-engineering. I'm just going in to see what they say."

"So you have no idea what you would be doing?"

"No idea. It just popped up on my opportunity queue.

It said I have to go in and take a bunch of tests and then they would see if I qualify for a series of interviews."

"A *series* of interviews?"

"I know; it's ridiculous. But it's also what makes me think it might be something important.

They don't specify anything. I don't even know what they're going to ask."

"Maybe they want it that way. Maybe they don't want people to prepare."

"Hopefully that means I've got as much of a chance as anyone else. I say that, and I don't even know what the job is."

"Maybe it will be something cool. Maybe they're wanting civilian astronauts or something."

"You're just hoping they'll shoot me out into space so you can be done with me."

"I was thinking more along the lines that I've never done it with an astronaut."

"I would hope there's a lot of occupations you haven't done it with. You make it sound like there's only a couple left."

"Don't worry, you're totally in my top seven."

"Nice. Real nice."

"Geo-engineering. I know they have stats on us and all. But why would they have picked you for this?"

"Who knows? The only reason why I'm considering it is that if I get it, it's saying we can live on campus."

"In the Zones?"

"Yep."

"I thought the Zones was like this thing where once you go in, you can't go out."

"Just depends on what kind of clearance you have. A buddy from school got a job doing low level stuff and doesn't even live on campus."

"Are you excited?"

"I'm excited about the idea of getting us out of the Stretch."

"Yeah, but I mean more than that. This is kind of a big deal, right?"

"I guess it could be."

"What's wrong?" Annasophia no longer saw the smile she had earlier detected.

"I don't know. It's like there's a part of me that doesn't want to get my hopes up because I don't know what the job is. And even if it's cool, I don't know if I'll get it.

But then there's another part of me that felt, when I got the notification, that I had this chance to finally do something.

Before my dad died, he told me this story about how my grandfather sat him down when he was a kid and told him that when he got older and got to his age there wouldn't be any more jobs left like the one he had. And that when he did get older he realized my grandfather was right, but maybe for different reasons than he thought at the time. And now that my father realized that, he found himself having to tell me the same thing - that when I reached his age, there wouldn't be jobs like he had.

And I think he was right too, but again not for the same reasons he thought at the time.

I know we've struggled, but I also know we've got it better in the Stretch than a lot of other folks. But it seems like the only thing we really have to look forward to is knowing we were successful at running from a worse situation.

For the longest time, I've just resigned myself to thinking that the rest of my life is going to be whatever was given to me. And in the end, it's not going to matter. My life is going to be just me making sure I, or we, don't fall further down.

And I know it's about the whole having a good attitude thing and treating people good and all that, but it's not about all that. It's about mattering. And I feel like if all I do is make sure we don't fall further down, then what kind of life is that? Is that all we have? To try to hang on and not become as miserable as the people in the Tunnels or the scavengers in the Heaps? Or *if* we're lucky, and we can sell enough of ourselves that we can eventually have a chance at getting into the Perimeter; is that a life?

I've never had work that I felt good about, that I felt mattered even one bit. I'm serious. I've never had one thing show up in my opportunity queue that didn't look like it could be filled just as fine, if not better, by almost everyone I know.

When I was a kid and they had us take those tests for how we would fare after the Changes, I scored so high I thought I had it made. As soon as I saw the scores, I immediately started daydreaming about what my life would be like. I remember running home so I could tell my dad how well I did on the tests.

And I remember after I told him, all he said was, 'Would you rub my arms?' Because his arms were hurting.

At first, I thought it was his way of telling me not to have these unrealistic expectations. And maybe it was. But I also feel like he was letting me know that, even if I was to make it, it wouldn't change how much someone like him would still be hurting. There would still be all these people hurting.

So when I saw that this was in the Zones and it was part of one of the geo-engineering programs, I thought that just maybe I had a shot at playing even the tiniest part in stopping or at least lessening some of that hurt. And for a second, I felt like I did when I was a kid. And it was like I was almost surprised that I was still capable of hope.

Except I don't want to feel hope. Because I know that I'm probably one of a few hundred people that got the same message and that are going to go in and take these tests, only to wait for the notification that we didn't make the cut.

And when that happens, I can't bear going back to that place where I just don't have any hope.

And don't get me wrong. You make me happy. And trying to make you happy gives me a feeling of purpose, maybe the only feeling of purpose I have.

But our happiness is in spite of what's happening to us. And it can't just be about that.

I've never been one of these people that wanted to pick up a gun and go live under a rock like the Resisters. But I get why they're doing it. Because feeling helpless is worse than feeling afraid.

So I guess... yeah, I'm excited, but it's like my excitement is being regulated by all these other thoughts."

"What were you saying? I kind of trailed off when you started talking about your feelings."

Richard tried his best not to smile, unsuccessfully. "I would say you're the worst, but it's kind of beyond measurement now."

"I'm just saying, you got a little deep on me."

"Don't say it."

"But it was kind of cute."

"That's it. You've done your duty. You can go back to hell now and collect your reward."

"No, but really. I get everything you're saying, and we've talked about this before. But the fact is you don't know what's going to happen, and all you can do is go in and see. And if there is a real shot at it turning out great, it's like you said. That's a lot more chance than we have right now."

"And I know it sounds hokey, but I was thinking that maybe this is some kind of sign or something."

"What kind of sign?"

"I was thinking that maybe this means something good really is going to come out of the geo-engineering programs. And if it does, then maybe we could revisit that conversation about having a baby."

"Are you serious?"

"Yeah. I mean, again, I don't even know what the job is, much less if I'll get it. But if I were to get it, I might be able to find out how much progress they're actually making. And if it looks like there's a chance of them figuring it out, then maybe I would feel confident in giving it a shot."

"And worst case, we would be living in the Zones. And you know kids have a good life in the Zones."

"You're right.

So anyway, I'm not saying that any of that's going to happen or that I'm even going to get past the first round of tests. I'm just saying it's the first time in a long time I haven't felt helpless."

"Well, ok then. Fingers crossed."

"Fingers crossed."

A way back to that place

The table and the chairs had been brought in. So had the china and glassware. And the paintings on the wall. There was scarce room for anything else.

She had chosen this room for its lighting.

It was not unlike the set in her own office. Everything had to be perfect.

This wasn't just any interview. This was a discernible moment, reserved for history.

A selection of music played as they ate. Songs from their shared years.

The food was also from a memory. The time she worked catering that banquet and they let everyone take home all those leftovers. That was the first time she had tried sushi. Goat cheese too.

The table was long. Galya had placed them at the respective heads. Had she been hosting, her guest would not have chosen the same.

Nevertheless, her guest enjoyed the care taken. Galya enjoyed the company.

For this, both found it easier, to not let on.

Although she knew Galya had not prepared the meal, her guest felt it no less appropriate to pay compliment.

"This is delicious. I had forgotten what the real thing tastes like. At home, we've convinced ourselves you can't tell the difference anymore. But with this, you can. As good as we've been able to make it, there's a distinct quality that just can't be replicated."

"Perhaps it's like varying degrees of digital audio."

"Oh, I remember those conversations. I hope I don't come off as pretentious as some of my self-described audiophile friends growing up."

"You mean Jared and Keith." Galya hadn't thought of them in years.

"Yes, Jared and Keith, sworn enemies of the MP3."

"You couldn't play anything without them going on about how awful the sound was."

"Except for vinyl."

"Except for vinyl. They only made exceptions for vinyl."

"That's so funny. I had forgotten for a second that we share those memories."

"17 years."

Her guest contemplated just how long it had been since she was 17, since they were the same person. "It's a bit strange reminiscing with yourself. Different from old friends, don't you think?"

"I suppose so, if just for the fact that we don't have any secrets."

"From those years.

I can't even imagine telling you some of the foolish things I did after that time. I think it would be harder telling you than anyone else."

"We usually carry more guilt for our actions than surprise for others."

"True. And yet I think if we opened that door, we'd be here 'til dawn."

Galya was careful to keep smiling. "You're probably right."

"Maybe another time." Her guest raised her glass.

Galya took cue. "To old times."

"To old times."

After taking her drink, Galya put her glass down and proceeded, this time with something more than small talk.

"Why did you also choose not to have a family?"

"I'm not sure if I would wholly characterize that part of my life as a choice. I did, however, presume, with what you accomplished and what I was left to accomplish, that I might be more *efficient* following your lead."

Galya had not known what to expect in her guest's answer. She turned back to small talk. "So what did you think of the facility?"

"It's awe-inspiring. And I should say, Richard is quite the tour guide."

"Yes, we were certainly fortunate to find him."

"I was impressed with how self-contained the facility is. The way you've embedded its sustaining features into the security apparatus is remarkable."

"With the unpredictable nature of things, the integrity of the structure necessarily requires heavy investment."

"Certainly. And we're hoping that we can substantially lessen that kind of investment with the advancements you helped us leap forward to. I know that Richard told you we've prepared a data drive with all our technology, and I think for the most part it's pretty intuitive. But there are a few areas in which I thought it might be helpful if we go over."

"Actually, I was hoping we could talk about something else."

"Of course. I'm so sorry." Her guest looked embarrassed. "You're gracious enough to have me as your guest, and I discourteously start talking shop. Forgive me."

"Not at all. As a scientist, I can't wait to analyze the progress you've made.

But there's something else."

Galya took another drink, to illustrate the casual nature she wished the conversation to maintain.

"From what Richard informed me, you've lived your life in service to our world as much as your own. For that, I owe you being frank in our discussion."

"Of course."

"As fascinating and satisfying as it will be for us to dig into your technological achievements, you should know our aim was not merely to bring back a better understanding of the physical world."

"Then what was it?" Though warranted, her surprise had a moderated tenor.

"The truth of the matter is, in our current state, we are past mitigation. We are even past adaptation. The speed in which our planet is becoming uninhabitable is too great for any scientific advancements. While the planet may eventually self-correct, the species won't make it across that bridge."

Galya sat up slightly and put her hands in her lap.

"We didn't send Richard to find new technology. We sent him to find a new home."

"I don't think I understand. From what Richard informed me, there are still billions of people in your world. Even with our grasp of sustainable inter-dependency, that kind of excess consumption would be too much for us to take on. It's an impossibility."

"And we don't expect you to take on that kind of burden.

The volume of guests we would like you to host is proportionally quite modest."

"Modest."

"Yes, it's what we believe is a reasonable request. We have taken into account your resources and instituted a selection process in order to minimize any disruption. Guests we believe would easily fit in and contribute."

Her guest sat up slightly and put her hands in her lap. "Am I to assume we would not have a say in that selection process?"

"We're offering the most highly educated, most accomplished, most advanced among us."

"And am I also to assume that you're not really asking?"

"I'm asking you to help us in crafting our communication with your people. We have a team working on the best possible scenarios, but we would be grateful for your input.

Again, to ensure a minimization of disruption."

"Something more subtle than 'make room or else'?"

Galya wasn't certain if the comment was meant to be light.

"As I stated, we would be grateful for your input."

"Is this why you agreed to meet me?"

"It was a practical consideration. But I would be lying if I said that the personal didn't play a part."

"And I assume you're not allowing me to go home."

"You can go home once we're home."

"And what if I do help you and you make this communication, and my people say no?"

"Whenever we create new coordinates, we send others to confirm what's been reported, while also assessing the defense capabilities of our potential host."

Her guest nodded. "How poetic. You went searching for us just so you could go to war with us."

"We didn't search for you." Galya couldn't let the comment go unaddressed.

"We created you."

"Then you created us so you could go to war with us."

"We created a home for our people."

"For some of your people."

Galya took no offense at the implication.

"As you said, we couldn't expect you to take everyone. It would be unreasonable."

Her guest took a drink, to illustrate the casual nature she wished the conversation to maintain.

"Did Richard know what he was doing?"

"There was no need to tell him. There was no room in his imagination for the truth, and no room in his conscience for certain questions. We choose our heroes very carefully."

As her guest returned the glass to its place on the table, she ran her finger down and up the refinement of the crystal, admiring the shine. This took about seven seconds.

Then she spoke. "You know why we got it right?"

She withdrew her hand from the glass.

"It's because Richard showed me how I failed in his previous visits. The real tools were not in the technical knowledge he provided, but in examining why I had failed, how I could have failed, with such advanced gifts.

I recognized that each time I was under the same impression that he was: that it was just about the technology, that earlier implementation would be enough. And then I realized the reason I kept failing is the same reason you're in the place you are right now.

The magnitude of what was given to that 17-year-old girl working in a diner was overwhelming. Still, Richard thought I was the right person to have control over it. And while that possession was enchanting, I studied the contents of that gift through a lens of self-doubt.

Knowing how many times I had tried it before, I began to question if Richard had made the right decision. And then I began to wonder if the problem wasn't that I had been controlling it, but that I was allowed to control it. Something that important.

I knew from our shared history that as long as something that important could be captured and guarded, there would be perpetual struggle to either wrestle it away or to make it irrelevant. And that contest would always poison even the best of our intentions and cause us to fall short of our potential. I knew the kind of power that followed this technology would inevitably lead to a certain paranoia, and paranoia has a way of framing the imagination.

I knew, even with the information I had in that folder, I couldn't be trusted with it. So I did away with the ability to keep it. To keep it for oneself. To keep it from others.

I took those lottery winnings, and I spent it on figuring out a way to make all my money worthless.

Because even with those hundreds of millions of dollars, I knew it wouldn't be enough. This society was crafted to defend against its own salvation.

Luckily, that defense depended on having someone to attack. And I really had no more idea what the future was supposed to look like than anyone else. I just knew what some of the pieces looked like.

So I began to give away those pieces. To the people, to public domain. I did it over time, often cryptonymously. And little by little, strangers began working together, putting together a puzzle. Granted, they didn't know they were putting together a puzzle. But that didn't matter. What mattered was that it was theirs to put together. Not mine.

And as I released each piece of that puzzle, it reconfigured the access and flow of technological interaction so that it incentivized mass participation, so that the direction of its innovation was a matter of consensus rather than the perceived spoils of perceived battles.

And as this process enabled new possibilities of political and economic vision, the structure and the culture of our failings crumbled away. It wasn't that the technology had changed. It was that the people had changed.

It was no longer my experiment. It was theirs."

"That's the nature of creation. It's ultimately about losing control.

Richard was right to give you the information. He just didn't know why he was right."

"Just like you don't know."

"Why do you say that?" It had been a long time since anyone had told Galya she didn't know something.

"This isn't about your people being saved, is it?

It's about you saving them.

I know my capacity for ego and vanity. Just like I knew the longer I was in that world of limited imagination, it would corrupt me. And I would eventually become a bureaucrat, content to implement the grand designs of gatekeepers - in return for a title and some measure of worship."

As she spoke, her tone became softer, but still firm.

"You don't have to do this. You don't have to be this.

We re-envisioned our society, and we can show you how to re-envision yours. Not just for a few, but for everyone.

We believe that our technology *can* save you, but only if it's employed under a radically different construction of human interaction. Once I realized that we would avoid the fate Richard described, I put together a team of innovators to prepare for our reunion.

It took us over thirty years to do it, but we have an implementation model for every possible environmental outcome. We have a model for your outcome.

It's all right there in the data drive I provided. I can show you how to implement it. That can be your new home.

You just have to accept it."

Galya smiled affectionately.

"I'm sure this will be a help to those who remain." She then resumed her meal.

Her guest refrained from joining her, allowing Galya a few bites before interrupting.

"You know we weren't the first to find a balance with our resources. It's happened before. Many times in history. Different places, different peoples. Each with entirely different understandings of the world around them. Each coming to a place of balance.

And yet none of them were able to keep it.

When gifted with enough weapons, diseases, or chains, each society watched that balance turn to dust.

And yet it wasn't a rival's technology that did them in. It was enough children on the playground believing that the bigger, stronger kid had won the argument.

The kind of violence required for that kind of victory is not in its technology. It's in telling the defeated, enough times, that this lust is somehow natural. That the taking is a natural order.

That's what true violence looks like. It doesn't just kill people. It convinces them. It reduces their renaissance to a novelty and their wisdom to a mistake.

We didn't escape your fate because we evolved. We escaped your fate because we threw off what we thought we knew and found a way back to that place, where humans had once been, where they had already evolved. And we refused to be the people who killed them.

We abandoned that thirst. We shed those tools. That was our answer.

Your answer, however, was to set sail again, finding a new place to start it all over."

"Again, we did not discover your world. We created it."

"You created a place that could never be your home. Our reality looks nothing like yours. Our economy. Our governance. Our sense of community. You would only resist our way of life. You would only try to change it.

You don't want what we have. You want what we vowed to never have."

Galya was starting to enjoy this, back and forth. It was entertaining. It was a surprise.

"Maybe you're right. Maybe we would hate your way of life. Maybe we would try to change it. But even if that's true, isn't what you've built strong enough to guard against that?"

"We already won those battles. And it took decades.

What you see as education, we see as delusion. What you see as accomplishment, we see as a placation, mollification.

You think you're giving us the best of your people. You're giving us your worst.

How could we agree to that?"

"It's not about your agreement. It's about what you owe."

Her guest looked down, in resignation, at what little was left of her food.

"I didn't want to believe them.

When I put together my team, there were those who suggested this might be a possibility. That we might need to be prepared for the possibility that Richard had lied to me or that he was being lied to.

It took me years to even entertain the idea. But eventually I acknowledged that if it was even a possibility, and the world we were building was to mean anything, there would have to exist a plan to defend it."

Galya tried very little to help herself from smirking. "To defend it."

"Of course the science was there. But still. Having outgrown our need for weapons of war, the resources we invested into these concerns were a source of shame."

"If you're suggesting what I think you're suggesting, you should know that whatever you have prepared won't work. We can send entire armies wherever we want and recall them in a second."

"That of course was the hardest part. Coming up with the multitude of scenarios. We didn't even know if Richard would be the one to come back. And if it was him, we didn't know if he would contact me.

We certainly didn't know if you would allow me this visit. That's why we didn't tell Richard. We couldn't risk his reaction.

He had to come back. In case."

"Is this really your attempt to scare me?" Galya was no longer entertained. "Because if it is, it's pathetic. Richard has no access, no information. He's an order taker who's under surveillance at all times, just like the rest of his team. And just like the rest of them, now that we've found you, he's no longer needed."

"We call them smart viruses: selective, programmable, undetectable."

"I'm not sure whether you think this is funny or if you're just desperate and trying to bluff me." Disgusted that her courtesy had been taken for granted, she was nearing a decision to conclude the dinner.

"They're as horrific as they are impressive. The one we gave..."

"Had I thought you'd react in such a childish manner, I wouldn't have told you at all." Galya could no more process such a conclusion than she could the fact that she was no longer in control of the conversation.

"The one we gave Richard is an airborne variant. It would have taken a few extra days and, with your security infrastructure, would have only spread through the compound before it died out. The one I brought is also airborne, but a bit different. It works on a..."

Finally, Galya shouted, "I can give the order, and a guard will come into the room and put a bullet in your head!"

"That's assuming they're not already asleep.

Having a reliable model for your DNA, I asked if you could have a little more time than the rest."

Galya could feel her heart rate increasing. She tried to compose herself as she unlocked her communication device and pressed for assistance.

"Guard, can you come in here?"

"I'm glad that Richard didn't know. It gives me hope. And will make our plans easier."

"Guard, come in here right now."

"Don't worry. It doesn't hurt. It just puts you to sleep."

Galya checked the visual on her communication device.

"What did you do?"

"This really is a good meal; we might as well enjoy it."

Her guest smiled affectionately and resumed her eating.

Noticing that Galya was refraining from joining her, she took a few more bites and then assured her.

"I promise it's not going to hurt. You may just feel a little weak beforehand."

"How could you do this? How could you not save us?"

"Oh, but it was you who first suggested that not everyone could be saved. I think you called it a selection."

"That is not what I meant."

"Do you know what convinced me that you had not come for technology? It was Richard alluding to how many times he had come up short, and that he wasn't the only one assigned this task.

That there were others doing the same thing. This trial and error. Each one like Richard, simply an employee, going to work day in and day out. Thinking the worst thing that can happen is a fruitless branch.

I don't know what it looks like beyond the gates here. I imagine the suffering is quite immense.

And yet how many of these branches have you created? Hundreds? Each one host to billions more, billions more doomed to the same life as those beyond your gates.

You've taken your own folly and the suffering it commanded, and you've made it over and over and over.

And now you demand that we provide a home for a handful of your educated and accomplished.

Where is the new home for all those other timelines, all those realities, all those suffering beyond the gates?"

"They had the same chance to save themselves as we did."

Her guest took another bite of her food.

"They say that you can learn any language if you're immersed in it long enough. But I have to say, it's quite jarring to hear your own voice speak in one so foreign."

"Then show us. Save us. Give us a chance to learn your language.

If you believe we are sick, then heal us."

"It's not that we believe you could not learn our language. It's that we cannot allow yours to be spoken again."

"Do you not owe us anything?"

"Maybe. But how much?
Do we owe you enough to risk being destroyed by you?"

"Then save me. Surely you can take one person.
Do you not at least owe me that?" Coming to terms with the next few minutes, Galya couldn't see the irony holding together her incredulity.

"It's correct that we owe you.
But we don't owe you that."

"I created you."

"You created us, then clothed us with your debt."

"Is that what you call your existence, a debt?
My God! Are all creations this ungrateful?
Can children be this spoiled, that life itself is not enough?"

"I'm not your child."

"But you are my creation," Galya called across the table.

"You damned us by creating us."

"But still I am your creator."

"A creator in need of salvation."

"Then save me." It was less an imperative, than a plea.

"I am saving you. By not allowing you to do this to anyone else.

Even if we could hold you off, it wouldn't stop you. You would simply send Richard, and those like Richard, back to work.

That cannot continue. That's why I came. That's what I owe you."

"And what about my work, what about this technology?"

"You mean these secrets?

In my experience, people don't miss what they don't know exists.

You murdered your discovery the moment you decided to keep it to yourself."

Galya sat up again, straight. "If you destroy all this, there's no way you can get out of here, no way you can get back home.

Are you willing to let yourself die?"

Her guest took a drink.

"That's a possibility I had to grapple with.

And maybe it's just a philosophical exercise, or maybe I'm superstitious. But I can't help feeling that we are linked. That what you have done is what I have done. And that I too am responsible."

"And this is responsible?

You think letting us die is the moral choice?"

"As opposed to your choice?

You had the choice to stop this. After the first failure, you could have put it to an end. Yet you chose to continue. Each failure, one after the other. You couldn't be bothered with their suffering. As long as you believed that a better world was within your reach, you would allow this to happen.

And now it has happened, and we believe a world absent of this madness is the best world within our reach."

"I know what you think of me. But you only know 17 years. You don't know what I had to do to get here, what I had to sacrifice. Only to then be informed what my work would be used for.

Does it matter to you that I was not given that choice? Do you not see the irony that you are now not giving me a choice?"

"You don't have to do this."

"I'm telling you the truth. This is not what I wanted."

"Maybe it isn't. Maybe you didn't want any of this. Maybe you thought this was the best of your options.

Or maybe you're just trying to save your life.

It really doesn't matter. Because spending even one moment trying to decipher the truth would make all of this - about you."

"But it is about me. Right now.

I can't leave like this. I'm more than dying like this.

What can I do? What can I do to convince you, to change your mind?"

"I wish I knew what to say. To make this easier for you."

"You can't deny it." Galya was almost crying. "You can't deny that I saved humanity. I did.
Is that not enough?"

"For whom? For us or for you?
If that is what you need, then isn't it enough?"

"Please, don't do this to me."

"I'm doing this with you." Her guest took her last bite.

"You still don't understand. I didn't come here to defeat you. I came here to make good on your promise of finding a better world. But in order to do that, I have to pay for our actions."

She then took the napkin from her lap and placed it on the plate.

"And now, if you'll excuse me, I have some work to do."

"Is there really a plan for us on that drive?"

"Of course."

"And you knew I wouldn't accept it."

"We assumed that to be the most probable outcome."

"Then why? If you assumed the worst of us, why spend all those years working on it? Why not spare yourself the trouble and just kill us?"

"The plan was not for you, not *now*.

The plan is for, as you said, those who remain."

Galya slumped over, weary. Her guest had the urge to comfort her.

But there was work to be done.

"Stop recording." Her guest picked up the heart from her watch, from where it had been lying on the table.

"Thank you for meeting me."

Galya felt she could slip out of consciousness at any moment.

"You disappoint me."

Her guest understood, and she meant what she was about to say.

"I'm sorry."

She then went to the door, placed her hand on the sensor, and walked out.

She figured Galya had another minute or two. She thought it best to leave her alone, to think.

Galya didn't know, in this moment, what there was, to think. So she revisited what she had told herself before the dinner began. That it was ok. Because she was the original.

www.ingramcontent.com/pod-product-compliance
Lightning Source LLC
LaVergne TN
LVHW040515210926
848480LV00005BA/1467
* 9 7 9 8 9 8 5 5 8 3 0 4 5 *